Computer-Aided Engineering Design with SolidWorks

Computer-Aided Engineering Design with SolidWorks

Godfrey Onwubolu

Humber Institute of Technology & Advanced Learning, Canada

Imperial College Press

ICP

Published by

Imperial College Press
57 Shelton Street
Covent Garden
London WC2H 9HE

Distributed by

World Scientific Publishing Co. Pte. Ltd.
5 Toh Tuck Link, Singapore 596224
USA office: 27 Warren Street, Suite 401-402, Hackensack, NJ 07601
UK office: 57 Shelton Street, Covent Garden, London WC2H 9HE

British Library Cataloguing-in-Publication Data
A catalogue record for this book is available from the British Library.

ISBN 978-1-84816-665-3

Printed in Singapore by B & Jo Enterprise Pte Ltd

Preface

This textbook is designed for students taking the SolidWorks course in colleges and universities, and for engineering designers interested in using SolidWorks for real-life applications in manufacturing processes, mechanical systems, and engineering analysis. The course material is divided into two parts. Part I covers the principles of SolidWorks: simple and advanced-part modeling, assembly modeling, drawing, configuration and design tables, and surface modeling. Part II is unique because this book is one of the very few that covers, to any great extent, the applications of SolidWorks. Most other textbooks cover mainly the content of Part I of this book with only one or two applications.

The sections on manufacturing processes include the design of molds, sheet metal parts, dies, and weldments. The sections on mechanical systems include aspects of routing such as piping and tubing, gears, pulleys and chains, and cams and springs. They also cover the design and analysis of mechanisms, as well as threads and fasteners. The sections on engineering analysis also include finite element analysis.

This textbook is written using a hands-on approach in which students can follow the steps described in each chapter to model parts, assemble parts, and produce drawings. They will be able to create applications on their own, with little assistance from their instructors during each teaching session or in the computer laboratory.

This book has a significant number of pictorial descriptions of the steps that a student must follow. There are lots of students' projects and audio-visual clips covering most chapters in Part I. Instructional support is also provided, including SolidWorks files for all models (see Resources for Readers, page xi), drawings, applications, and answers to end-of-chapter questions.

The principles and exercises presented in Part I have been tested in the introductory and advanced SolidWorks courses that the author taught in Toronto, Canada. All the examples in this book have been solved by the author.

Dedication

This book is dedicated entirely to God who did *solid works* in creation of everyone and everything in existence and in sustaining life. I owe Him all that I have because all that I have comes from Him.

Acknowledgments

The author is grateful to a number of people who made the publication of this book a reality. The author was first exposed to computer-aided engineering design when he worked on his doctoral thesis at Aston University, Birmingham, England, when he developed a geometric modeling system to perform engineering functions similar to the kernel on which today's off-the-shelf geometric modeling software, such as SolidWorks, CATIA, Pro/E, and Inventor, are based. Most users of these software programs use them as black-box tools, without knowing how they function internally. Therefore, my research work at Aston University introduced me to this exciting area. Yasser Elkady introduced me to SolidWorks; I appreciate the extent of his knowledge in SolidWorks. I appreciate my students whom I taught SolidWorks and who gave me much feedback on the content of this textbook and worked hard on a number of projects that I proposed to them, and supervised until they were successfully completed. Sarah Haynes was the Editor whom I worked with initially before she left Imperial College Press. I owe much appreciation to Lance Sucharov, the Commissioning Editor, and Catharina Weijman, the current Editor, for their patience and assistance in getting this book project to a successful completion stage.

My wife, Ngozi, and our children are greatly appreciated. My wife, Ngozi, shared some very challenging times with me during the early stage of my learning SolidWorks and when I contemplated writing this textbook. Without the role that members of my family played, this textbook project would not have succeeded.

Godfrey Onwubolu
Toronto, Canada
January 2013

About the Author

Dr. Godfrey Onwubolu currently teaches **computer-aided design (CAD) using SolidWorks**, and **engineering analysis using SolidWorks** as well as *Applied Mechanics* and *Mechanics of Materials* where he applies SolidWorks to both teaching and applied research here in Toronto, Canada. He holds a BEng degree in mechanical engineering, and both an MSc and PhD from Aston University, Birmingham, England, where he first developed a geometric modeling system for his graduate studies. He worked in a number of manufacturing companies in the West Midlands, England, and he was a professor of manufacturing engineering, having taught courses in design and manufacturing for several years.

He has published several books with international publishing companies, such as Imperial College Press, Elsevier, and Springer-Verlag, and has published over 130 articles in international journals. He is an active Senior Member of both the American Society of Manufacturing Engineers (ASMfgE) and the American Institute of Industrial Engineers (IIE).

Recommended Reading

Engineering Design and Graphics with SolidWorks, Bethune, J. D. Prentice-Hall, 2009.

Howard, W. E., and Musto, J. C, Solid Modeling Using SolidWorks 2008, McGraw-Hill Ryerson, 2008.

Lombard, M., SolidWorks 2009, Wiley Publishing Inc, 2009.

Planchard, D. C., and M. P. Planchard, A Commands Guide for SolidWorks 2008, Thomson/Delmar Learning, 2008.

Planchard, D. C., and M. P. Planchard, Engineering Design with SolidWorks 2009, Schroff Development Corporation, 2009.

SolidWorks Tutorials, SolidWorks Corporation.

Tickoo, S., and Sandeep, D., SolidWorks 2009 for Designers, CADCIM Technologies, 2009.

Resources for Readers

SolidWorks resources designed to support users of this text have been made available for download from the publisher's website. These files are separated into those for users (i.e. for students) and those for instructors.

The files for users include of a number of SolidWorks exercises, allowing students to enjoy a hands-on learning experience with SolidWorks. They encourage students to follow the step-by-step descriptions given in each chapter, facilitating them in solving the problems presented in the text on their own.

The instructors' section provides full access to all the SolidWorks files used in the book, including solutions to the exercises and end-of-chapter problems; none of which are included with the users' files.

Please note that while the users' files are available to all, the instructors' files are restricted and will require an access code (available from the publisher) before they can be downloaded.

Users: http://www.worldscientific.com/worldscibooks/10.1142/p761

Instructors: http://www.worldscientific.com/worldscibooks/10.1142/p761-sm

Contents

Contents

Chapter 1

SolidWorks Overview

Objectives:

When you complete this chapter you will:

- Have a good background knowledge of SolidWorks
- Have learnt how to start a SolidWorks session
- Understand the SolidWorks user interface
- Have learnt how to set the document options
- Have learnt how to set up good file management
- Have learnt how to start a new document in SolidWorks
- Have learnt how to model your first part
- Know about useful SolidWorks resources

1.1 Background Knowledge about SolidWorks

SolidWorks, registered trademark of the Dessault Systems Corporation in Concord, Massachusetts, USA, is a design automation software package used to produce parts, assemblies, and drawings. It is a Windows native 3D solid-modeling computer-aided design (CAD) program based on *parametric modeling*. This particular attribute of SolidWorks means that the dimensions of the parts, assemblies, and drawings drive the shapes produced. SolidWorks provides easy-to-use, high-quality design software for engineers and designers for creating 3D parts, assemblies, and 2D drawings that are all related. This means that changes made in a part document will affect the assembly and 2D drawing documents. The advantages of parametric modeling are numerous, but some distinct ones include the fact that when a designer realizes that changes need to be made to some dimensions in an assembly document, making such changes in the part document automatically updates the assembly and 2D drawing documents. There is no need to start all over. Parametric modeling is not rigid; it is flexible and makes designing extremely flexible. This attribute of SolidWorks makes it extremely flexible when compared to some other CAD programs.

1.2 Starting a SolidWorks Session

There is more than one way to start a SolidWorks session. To start a SolidWorks session, choose Start > All Programs > SolidWorks from the Start menu or double-click on the SolidWorks icon on the desktop of your computer. When the SolidWorks program is first accessed, the window that is displayed is the one shown in Fig. 1-1. It is more or less a blank page from which the designer then decides what document to open: part, assembly, or drawing.

Figure 1-1 SolidWorks initial screen

1.3 SolidWorks User Interface

There are basically four menus that are first seen when a new SolidWorks session commences: menu-bar menu, menu-bar toolbar, SolidWorks help, and SolidWorks resources. The menus are visible when you move the mouse over or click the SolidWorks logo. You can pin the menus to keep them visible at all times. In order to create a part, assembly, or drawing, click File > New from the menu-bar menu or click New ⬜ or click Open 📂 (standard toolbar) from the menu-bar toolbar. The menu bar contains the following.

1.3.1 *Menu-bar toolbar*

This is a set of the most frequently used tool buttons from the standard toolbar as shown in Fig. 1-2. By clicking the down arrow next to a tool button, you can expand it to display a flyout menu with additional functions. When the cursor is moved across the SolidWorks logo, the menu-bar toolbar switches over to the menu-bar menu. The toolbar moves to the right when the menus are pinned.

Figure 1-2 Menu toolbar

The available tools are:

New ⬜▾   Creates a new document.

Open 📂▾   Opens an existing document.

Save 💾▾   Saves an active document. This lets you access most of the file menu commands from the toolbar. For example, the Save flyout menu includes Save, Save As, and Save All.

Print 🖨▾   Prints an active document.

Undo   Reverses the last action taken.

Rebuild   Rebuilds the active part, assembly, or drawing.

Options   Changes system options and add-ins for SolidWorks.

1.3.2 *Menu-bar menu*

The SolidWorks menus are visible when you move the mouse over or click the SolidWorks logo. You can pin the menus to keep them visible at all times. The default menu items for an active document are: File, Edit, View, Insert, Tools, Window, Help, and Pin (see Fig. 1-3). However, the menu items change depending on which type of document is active.

Figure 1-3 Menu-bar menu

1.3.3 *Task pane*

The task pane is displayed when the SolidWorks session commences. The task pane contains the following default tabs: SolidWorks resources , design library, file explorer, SolidWorks search, view palette, appearances/scenes, and custom properties.

SolidWorks Resources

The SolidWorks resources tab in the task pane includes commands, links, and information.

4

Design Library

The design library [icon] tab in the task pane provides a central location for reusable elements such as parts, assemblies, and sketches; but not for non-reusable elements such as SolidWorks drawings, text files, etc. It is the gateway to the **3D Content Central** website. The design library abounds with resources for design, and designers should take time to become familiar with these resources.

SolidWorks Explorer

The SolidWorks explorer [icon] is a file management tool designed to help you perform such tasks as renaming, replacing, and copying SolidWorks files. You can show a document's references, search for documents using a variety of criteria, and list all the places where a document is used. Renamed files are still available to those documents that reference them.

SolidWorks Search

The SolidWorks search [icon] is used to search key words. The ten most recent searches may be found by clicking the pull-down arrow.

View Palette

The view palette [icon] tab provides the ability to drag-and-drop drawing views of an active document, or click the Browse button to locate the desired document.

Appearances/Scene

The appearances/scene [icon] tab gives a useful and easy-to-use way of providing photo-realistic rendering of models.

Help options

The SolidWorks flyout menu of help options  is used to access help, tutorials, the reference guide, and other functions as shown in Fig. 1-4. Another route for help options is through the SolidWorks menu.

Figure 1-4 Help options

So far, we have considered the menus that are displayed when a new session of SolidWorks is started. Once a task begins (part modeling, assembly modeling, or drawing), three more menus appear: CommandManager, FeatureManager design tree, and the Head-up view toolbar. These are now briefly discussed.

1.3.4 *CommandManager*

The CommandManager, which is document dependent, contains most of the tools that you will need to create parts, assemblies, or drawings. The CommandManager is a context-sensitive toolbar that is dynamically updated based on the toolbar you want to access. By default, it has toolbars embedded in it based on the document type. For example, the default part tabs are Features, Sketch, Evaluate, DimXpert, and Office Products. These tabs are illustrated hereafter. The two most widely used categories of tools for part modeling are feature tools used to create and modify 3D features, and sketch tools used in creating 2D sketches.

The main feature tools used to create and modify 3D features are Extruded Boss/Base, Revolved Boss/Base, Swept Boss/Base, Lofted Boss/Base, and Boundary Boss/Base for adding materials to a part; Extruded Cut, Revolved Cut, Swept Cut, Lofted Cut, and Boundary Cut for removing materials from a part; Fillet, Linear Pattern, and Mirror for operations; specific features such as Rib, Draft, Shell, Wrap, Dome; Reference Geometry and Curves, as well as the Instant3D tool (see Fig. 1-5).

Figure 1-5 Feature tools

The sketch tools used in creating 2D sketches are Sketch; Smart Dimension, Line, Rectangle, Slot, Circle, Arc, Polygon, Spline, Ellipse, Fillet, Plane, Text, Point, Convert Entities, Offset, Mirror, Linear Pattern, Move, Display/Delete Relations, Repair Sketch, Quick Snaps, and Rapid Sketch (see Fig. 1-6).

Figure 1-6 Sketch tools

The evaluate tools are mainly for analysis such as measuring distances between points on features, mass properties, section properties, etc. SimulationXpress, FloXpress, DFMXpress, and DriveWorksXpress wizards are also accessible through the evaluate tools (see Fig. 1-7).

Figure 1-7 Evaluate tools

The DimXpert tools are mainly for dimensions and tolerance (see Fig. 1-8).

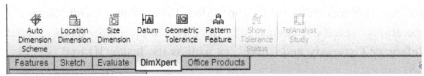
Figure 1-8 DimXpert tools

The Office Products toolbar allows you to activate any add-in application included in the SolidWorks Professional or Premium package. eDrawings and Animate are available using this toolbar as shown in Fig. 1-9.

Figure 1-9 Office Product tools

1.3.5 *FeatureManager design tree*

The FeatureManager design tree, which is located on the left-hand side of the SolidWorks graphics window, summarizes how the part, assembly or drawing is created. It is necessary to understand the FeatureManager design tree in order to be able to troubleshoot a model with problems. The default tabs of the FeatureManager are: FeatureManager design tree,

PropertyManager, ConfigurationManager, and DimXertManager. For example in the FeatureManager design tree shown in Fig. 1-10, the standard planes are shown, and the first activity is shown to be Extrude1 (extruded boss/base). A sketch (or sketches), which defines the extruded part, would normally appear when Extrude1 is expanded. All the details for designing a part are summarized and stored in the FeatureManager design tree. It is the design information storehouse, which fully describes a designed part, assembly, or drawing. It is a design tree showing the relationship between *parents* and *children* in a design context. When information relating to a parent is changed, it automatically affects the *children*. Every designer using SolidWorks should be very familiar with the semantics and syntax of the FeatureManager design tree.

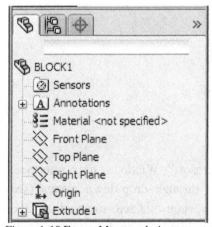

Figure 1-10 FeatureManager design tree

1.3.6 *Head-up view toolbar*

The Head-up view toolbar (see Fig. 1-11) is a useful tool for the user to view options during modeling of parts, assemblies, or drawings. The following views are available: Zoom to Fit, Zoom to Area, Previous View, and Section View, View Orientation, Display Style, Hide/Show Items, Edit Appearance, Apply Scene, and View Setting.

Figure 1-11 Head-up view toolbar

The view orientation (top, front, bottom, left, right, and back; isometric, trimetric, and diametric; normal; single view, two view horizontal, two view vertical, four view), and display style (shaded with edge, shaded, hidden lines removed, hidden lines visible, and wireframe) are illustrated in Fig. 1-12.

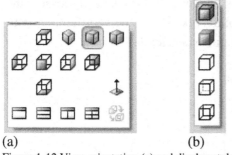

(a)   (b)

Figure 1-12 View orientation (a) and display style (b)

1.3.7 *Drop-down menu*

SolidWorks shares the familiar Microsoft® Windows user interface. Users communicate with SolidWorks through drop-down menus (see Fig. 1-13), context-sensitive toolbars, consolidated toolbars or the CommandManager tabs. As the name implies, a drop-down menu drops down other lower-level menus that give more information in terms of design toolbars. A drop-down menu has a black triangular-shaped symbol, which when clicked drops other menus. For example in Fig. 1-13, choosing the Features option, drops another menu showing the features available: Fillet/Round, Chamfer, Hole, Draft, Shell, Rib, Scale, Dome, Freeform, Shape, Deform, Indent, Flex, Wrap, etc.

Figure 1-13 Drop-down menus

1.3.8 *Right-click*

Right-clicking in the graphics window on a model or in the FeatureManager on a feature or sketch results in the display of a context-sensitive toolbar as shown in Fig. 1-14.

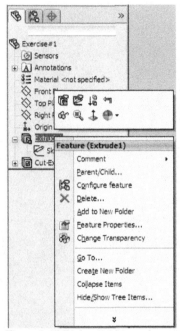

Figure 1-14 Right-clicking options

1.3.9 *Consolidated toolbar*

In the CommandManager, similar commands (instructions that inform SolidWorks to perform a task) are grouped together. For example, the family of the slot sketch tool is grouped together in a single flyout button as illustrated in Fig. 1-15.

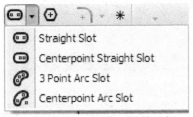

Figure 1-15 Group of the slot sketch

1.3.10 *System feedback*

SolidWorks provides system feedback by attaching a symbol to the mouse-pointer cursor that indicates what you are selecting or what the system is expecting you to select. Placing your cursor pointer across a model results in system feedback being displayed in the form of a symbol next to the cursor as illustrated.

1.4 Setting the Document Options

As has been already mentioned, the options tool changes system options and add-ins for SolidWorks. There are two components of options: system options and document properties.

1.4.1 *System options*

The system options tool allows you to specify file locations options, which includes a list of folders referenced during a SolidWorks session. The default templates folder for a new installation on a local drive *C:* is located at: *C:\Documents and Settings\All Users\Applications Data\ SolidWorks\SolidWorks200x\templates*. It is therefore important that you advise the SolidWorks software where to find your customized templates as shown in Fig. 1-16.

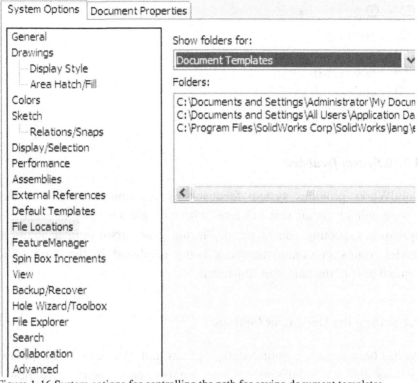

Figure 1-16 System options for controlling the path for saving document templates

1.4.2 *Document properties*

There are numerous settings that need to be made when you begin a modeling session. The Document Properties tool offers users the resources to do this.

Modifying the drafting standard

For example, to choose a drafting standard, the Drafting Standard tool is accessed. The standards that are available in this drop-down list are ANSI, ISO, DIN, JIS, BSI, GOST, and GB as shown in Fig. 1-17(a) and further expanded in Fig. 1-17(b).

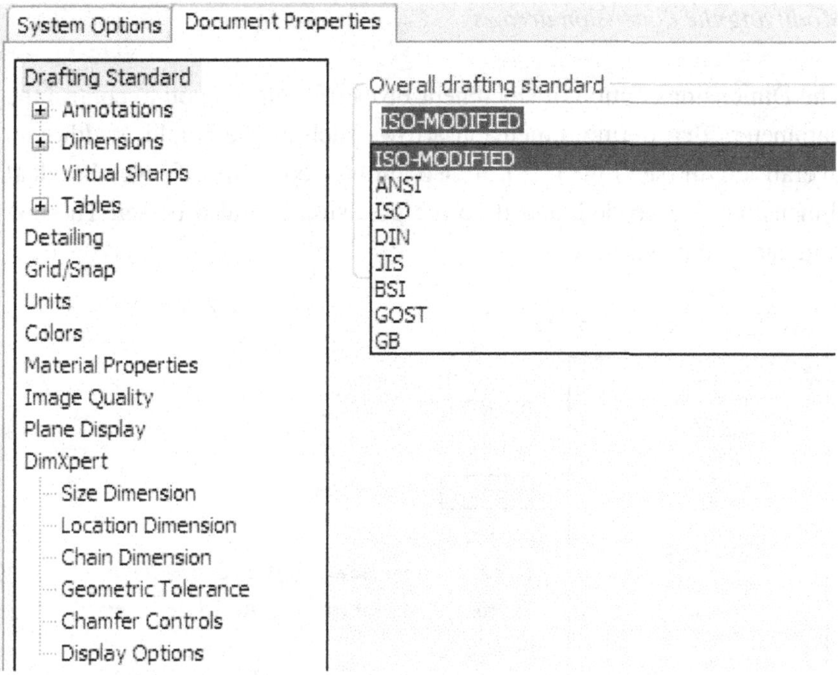

System Options	Document Properties	

```
Drafting Standard          ┌ Overall drafting standard ──────────
  ⊞ Annotations            │ ┌─────────────────┐
  ⊞ Dimensions             │ │ ISO-MODIFIED    │
     Virtual Sharps        │ ┌─────────────────────────────────┐
  ⊞ Tables                 │ │ ISO-MODIFIED                    │
Detailing                  │ │ ANSI                            │
Grid/Snap                  │ │ ISO                             │
Units                      │ │ DIN                             │
Colors                     │ │ JIS                             │
Material Properties        │ │ BSI                             │
Image Quality              │ │ GOST                            │
Plane Display              │ │ GB                              │
DimXpert                   │ └─────────────────────────────────┘
    Size Dimension
    Location Dimension
    Chain Dimension
    Geometric Tolerance
    Chamfer Controls
    Display Options
```

(a)

Drafting Standard Options	Description
ANSI	American National Standards Institute
ISO	International Standard Organization
DIN	Deutsche Institute fur Normumg (Germany)
JIS	Japanese Industry Standard
BSI	British Standards Institution
GOST	Gosndarstuennye State Standard (Russian)
GB	Guo Biao (Chinese)

(b)

Figure 1-17 Modifying the drafting standard

Modifying the dimension arrows

The Dimensions option in Document Properties allows you to control the parameters that define dimension arrows such as the height, width, and overall length (see Fig. 1-18). It also allows you to specify whether dual dimensions are needed, and if so the precision can also be set. The text font can also be specified.

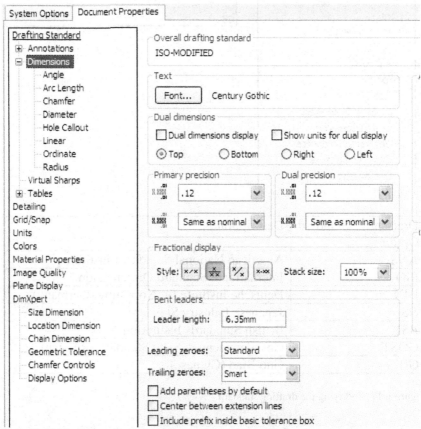

Figure 1-18 Modifying the dimension arrows

Modifying the units document properties

The Units document properties assist the designer in defining the Unit System, Length unit, Angular unit, Density unit, and Force unit of measurement for the Part document (see Fig. 1-19). The Decimals option displays the number of decimal places for the Length and Angular units of measurement.

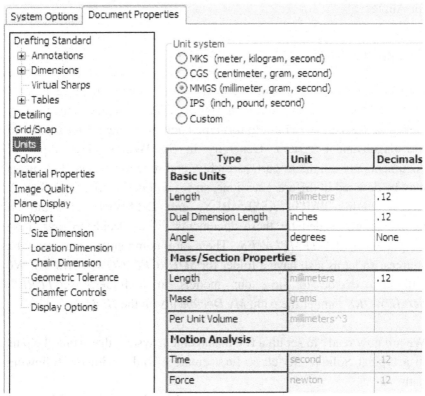

Figure 1-19 Document properties settings

There are hundreds of document properties. However, the document properties that are commonly modified include: Drafting Standards, Units and Decimal Places. For a number of applications we will need to set up the following:

ANSI-MM-PART TEMPLATE

Choose MMGS (millimeter, gram, second), decimal for length = 0.12 and None for Angle.

ANSI-INCH-PART TEMPLATE

Choose IPS (inch, pound, second), decimal for length = 0.12 and None for Angle.

1.5 File Management

An important aspect of working with SolidWorks is to have a well-organized file management system. Engineers and designers model many parts and it would be helpful to form the habit of having a well-organized file management system. Generally in SolidWorks, three types of documents are common: part, assembly, and drawing. These documents may be covered in different sessions, so we will define SESSIONS (1, 2, ...) or define PART, ASSEMBLY, and DRAWING. Customized templates are needed for these documents; let us decide to save these templates as *MY-TEMPLATES*. There are components that may be sources, so let us also have a folder for *SOURCED-COMPONENTS*. We will also decide to save our models in a folder, *LECTURES-SOLIDWORKS* under the path, *My Documents* on the *H:* drive.

We are now ready to set up a file management system that would help us in a typical SolidWorks class. First create a folder with the following path:

H:\My Documents\LECTURES-SOLIDWORKS

Under the *LECTURES-SOLIDWORKS* folder, create the following folder:
...\LECTURES-SOLIDWORKS\MY-TEMPLATES

Also, under the *LECTURES-SOLIDWORKS* folder, create the following folder:
...\LECTURES-SOLIDWORKS\SOURCED-COMPONENTS

Then under the *LECTURES-SOLIDWORKS* folder, create the following folders:
...\LECTURES-SOLIDWORKS\SESSION1
...\LECTURES-SOLIDWORKS\SESSION2
...\LECTURES-SOLIDWORKS\SESSION3

...\ LECTURES-SOLIDWORKS\ Session

Other folders could be created for projects, exercises, etc.

1.5.1 *Caution needed during a SolidWorks session*

During a SolidWorks session, it is necessary to save your model from time to time, because you may receive a surprise when your computer freezes. When this happens you might lose information that was not saved before the freeze.

1.6 Starting a New Document in SolidWorks

To select a new document (part, assembly, or drawing) in SolidWorks, select the New ▢ ▾ Document option from the menu bar. The SolidWorks screen of Fig. 1-1 should be open at this point. Another route to select New Document is through the Getting Started rollout of the SolidWorks resources. The New SolidWorks Document dialog is shown in Fig. 1-20. Click the Advanced button to select the advanced mode. The advanced mode remains selected for all new documents in the current SolidWorks session; the setting is saved when you exit the current session.

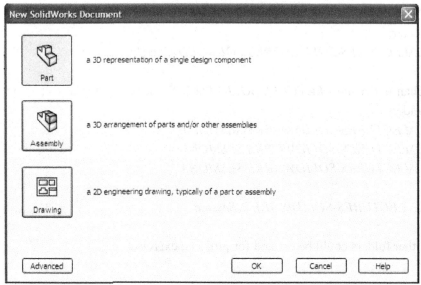

Figure 1-20 The New SolidWorks Document dialog

Part

The Part button is chosen by default in the New SolidWorks Document dialog. Choosing the OK button enables you to start a new part document to create solid models or sheet metal components.

Assembly

Choose the Assembly button and then the OK button from the New SolidWorks Document dialog to start a new assembly document.

Drawing

Choose the Drawing button and then the OK button from the New SolidWorks Document dialog to start a new drawing.

1.7 My First Part

After understanding what SolidWorks is all about, let us now create our first part.

Example 1

Create the model shown (left-hand side) in Fig. 1-21. A sketch of the model is shown (right-hand side). Ensure that the sketch is fully constrained. Extrude the sketch to a depth of 25 mm to produce the model. Change the extrusion depth and observe the model.

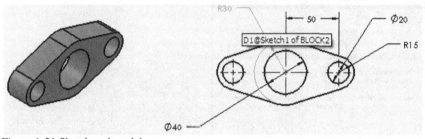

Figure 1-21 Sketch and model

Methodology for the solution

The following steps are recommended:

Create a New Document

Ensure that you have set up the following:

ANSI-MM-PART TEMPLATE: MMGS (millimeter, gram, second), decimal for length = 0.12 and None for Angle.

ANSI-INCH-PART TEMPLATE: IPS (inch, pound, second), decimal for length = 0.12 and None for Angle.

Choose the right-hand plane

At the origin, sketch a circle of 40 mm diameter

Sketch two circles of 20 mm diameter each

Sketch the outer profile: 4 arcs, four lines

Use relations to join the arcs and lines

Dimension the sketch

Extrude it 25 mm.

1.8 Useful SolidWorks Resources

The SolidWorks Tutorials tool (see Fig. 1-22), which can be accessed through the Help tool in the SolidWorks menu bar, is a very useful resource base. Explore the tutorials.

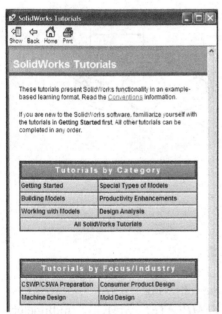

Figure 1-22 SolidWorks tutorials

1.9 Compatibility of SolidWorks with other Software

SolidWorks interfaces well with a number of standard CAD and application software (see Fig. 1-23). This means that while working with SolidWorks software, it is possible to import CAD files created using the software listed in Fig. 1-23.

Figure 1-23 SolidWorks interface with a number of standard CAD and application software

Exercises

1. Create the model shown (right-hand side) in Fig. 1-24. A sketch of the model is shown (left-hand side). Ensure that the sketch is fully constrained. Extrude the sketch to a depth of 15 mm to produce the model. Change the extrusion depth and observe the model.

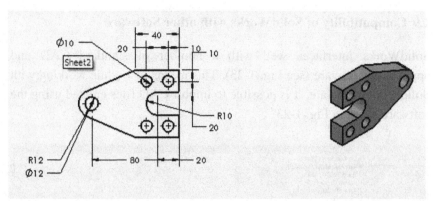

Figure 1-24: Sketch and model

2. Mirror the part about a plane that coincides with the two vertical rectangular faces.

Chapter 2

Sketch Entities and Tools

Objectives:

When you complete this chapter you will:
- Have learnt about sketch entities in SolidWorks
- Have learnt about sketch tools in SolidWorks
- Have used the sketch tools to modify sketch entities in order to produce parts

2.1 Introduction

SolidWorks sketch entities and tools can be accessed by clicking the Tools bar. These entities and tools facilitate the creation of parts using SolidWorks. While sketch entities have specific geometries, some of the sketch tools are used to modify the shapes of sketch entities.

Sketch entities include Line, Rectangle (two opposite vertices, center and vertex, 3 Point Corner, and 3 Point Center), Parallelogram, Slot (Straight, Centerpoint Straight, 3 Point Arc, and Centerpoint Arc), Polygon, Circle (center and one point, and three points on perimeter), Arc (center and two points, tangent, and 3 Point Arc), Ellipse (full, partial), Parabola, and Spline.

Sketch tools include Fillet, Chamfer, Offset, Convert Entities, Intersection curves, Trim, Extend, Split entities, Jog line, Construction Geometry, Make path, Mirror (straight, dynamic), Stretch Entities, Move Entities, Rotate Entities, Scale Entities, Copy Entities, and Pattern (linear, circular).

2.2 Sketch Entities

Sketch entities are commonly used geometric shapes, which are the building blocks for modeling simple and complicated shapes. In general, these entities are grouped into line, rectangle, polygon, slot, arc, circle, ellipse, parabola, and spline. Figure 2-1 shows the Sketch Entities toolbar.

For most of the entity definitions in this section, it assumed that a new Part has already been started, and the appropriate Plane chosen from the FeatureManager. Therefore, a generic format for sketch entities presented in this chapter is as follows.

Start a new Part document, click Sketch group on the Command-Manager, and select the (appropriate) Plane from the FeatureManager.

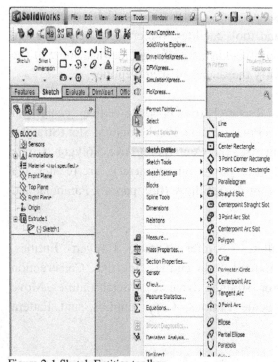

Figure 2-1 Sketch Entities toolbar

2.2.1 *Line*

Line is one of the most frequently used sketch entities in part modeling. There are basically two main types of line: solid lines and construction lines, as shown in Fig. 2-2.

1. Click the Line group on the CommandManager.
2. Select a starting point for the line and click there.
3. Select another point and extend the line.
4. When the line is complete, right-click the mouse and click the End chain option.

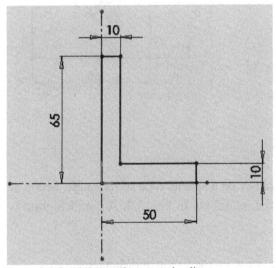

Figure 2-2 Solid lines and construction lines

2.2.2 *Rectangle*

There are a number of options for creating a rectangle entity: using two opposite vertices, the center and a vertex, 3 Point Corner, and 3 Point Center as shown in Fig. 2-3. When a user is more interested in centering the rectangle about the origin of the graphics screen, then the center rectangle is the one to go for. Let us illustrate the center rectangle.

1. Click the Rectangle group on the CommandManager.
2. Select the center and click it.
3. Drag the cursor away from the center point to create a rectangle.
4. When the rectangle is complete, click the OK checkbox on the PropertyManager to complete the rectangle construction.

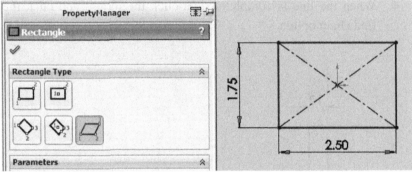

Figure 2-3 Rectangle entities

2.2.3 *Parallelogram*

A parallelogram is considered to be a special type of rectangle, hence, it is grouped as a rectangle type, as shown in Fig. 2-3. A parallelogram is shown in Fig. 2-4.

1. Click the Rectangle group on the CommandManager.
2. Select and click any two horizontal points.
3. Drag the cursor away from the second point to create a parallelogram.
4. When the parallelogram is complete, click the OK checkbox on the PropertyManager to complete the parallelogram construction.

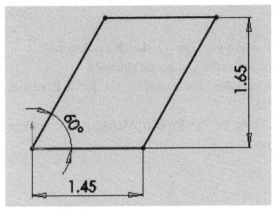

Figure 2-4 Parallelogram

2.2.4 *Slot*

There are a number of options for a slot entity: Straight, Centerpoint Straight, 3 Point Arc, and Centerpoint Arc as shown in Fig. 2-5.

1. Click the Slot group on the CommandManager.
2. Select and click any two horizontal points.
3. Drag the cursor to above or below the points to a third point, giving the width of the slot.
4. When the slot is complete, click the OK checkbox on the PropertyManager to complete sketching the slot.

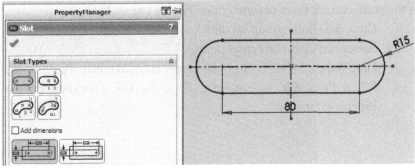

Figure 2-5 Slot entities

2.2.5 *Polygon*

A polygon has *n* sides, as shown in Fig. 2-6. To sketch a polygon:
1. Click the Polygon group on the CommandManager.
2. Define the number of sides, for example, six for a hexagon; check Inscribed circle.
3. Click the OK checkbox on the PropertyManager to complete sketching the polygon.

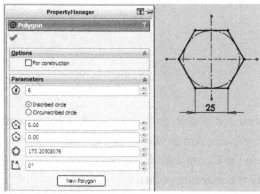

Figure 2-6 Polygon entities

2.2.6 *Circle*

There are generally two ways of defining a circle: the center and a point on the perimeter, or three perimeter points. See Fig. 2-7.
1. Click the Circle group on the CommandManager.
2. Select and click the center point.
3. Drag the cursor away from the center point to create a circle.
4. When the circle is complete, click the OK checkbox on the PropertyManager.

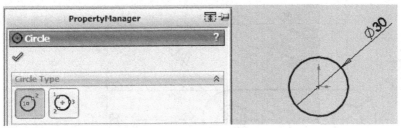

Figure 2-7 Circle entities

2.2.7 *Arc*

There are generally three ways of defining an arc: the center and two points, a tangent, or a 3 Point Arc. See Fig. 2-8.

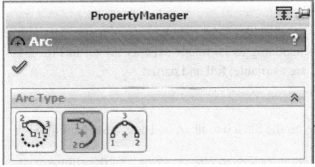

Figure 2-8 Arc entities

When the center point and two other points are known, the center and two-points option should be used. When two lines are already sketched and two end points of the arc are to coincide with the end points of the two lines, then tangent options should be used. In general, for an arc where the center point is not initially known, then the three-point arc option should be used. For example, in Fig. 2-9, only two points (end points of the lines) are defined for the tangent arc. However, the center and the two end points of the lines are needed when the center and two-point arc option is used, as in Fig. 2-9.

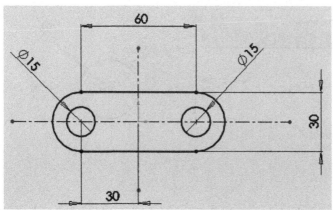

Figure 2-9 Arc entities

2.2.8 *Ellipse*

An ellipse has a major diameter corresponding to its major axis and a minor diameter corresponding to its minor axis, as shown in Fig. 2-10. Two types of ellipse are available: full and partial.
To create an ellipse:

1. Click Ellipse on the Sketch toolbar, or Tools > Sketch Entities > Ellipse.
2. Click in the graphics area to place the center of the ellipse.
3. Drag and click to set the major axis of the ellipse.
4. Drag and click again to set the minor axis of the ellipse.

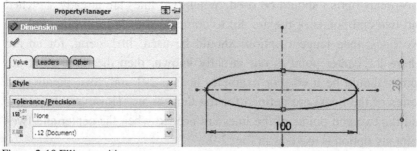

Figure 2-10 Ellipse entities

An ellipse can be used to model a rugby ball, by following these steps:
1. Trim half of the ellipse.
2. Sketch a line to close the shape.
3. Revolve the shape through the center line.

2.2.9 *Parabola*

A parabola is a loci of points in which the distance between a fixed point, the focus, and a fixed line, the directrix, are always equal. See Fig. 2-11. It is a well-known shape in elementary mathematics. To create a parabola:

1. Click Parabola (Sketch toolbar) or Tools > Sketch Entities > Parabola.
2. Click the origin and a point directly below it to place the focus of the parabola and drag to enlarge the parabola. The parabola is outlined.
3. Click on the parabola and drag to define the extent of the curve.

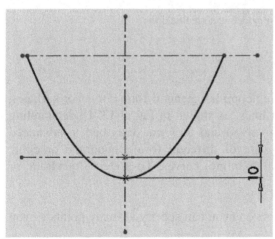

Figure 2-11 Parabola entity

A parabola can be used to model a satellite dish, by following these steps (see Fig. 2-12):

1. Trim half of the parabola.
2. Sketch a horizontal line at the top and a vertical line through the focus.
3. Revolve the shape through the vertical center line.
4. Shell the object with a thickness of 1 mm.

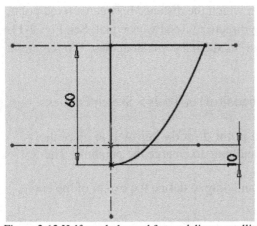

Figure 2-12 Half-parabola used for modeling a satellite dish

2.2.10 *Spline*

Splines are used for modeling complex, general-form curves or surfaces, through a series of straight lines, as shown in Fig. 2-13. Understanding the concept of splines is involved and they are described in advanced CAD documentation. They are of different forms: quadratic or cubic spline (non-uniform rational B-spline) curves. To sketch a quadratic or cubic spline, specify:

1. The first point.
2. The next point and so on (you can specify as many points as you want).
3. Complete the spline. The spline realized is a single entity.

The shapes modeled using splines can be "controlled", as shown, using controlling polygons. Splines are heavily used in the automotive and aircraft industries, and early work in the area began in those sectors for modeling automotive bodies and fuselages.

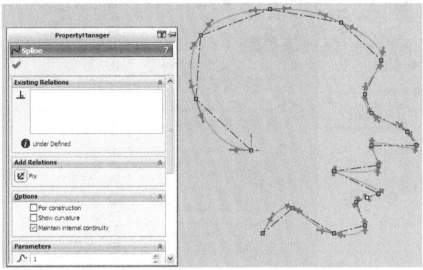

Figure 2-13 Spline entity

2.3 Sketch Tools

Sketches have a 2D nature and Fig. 2-14 shows the SolidWorks Sketch Tools toolbar mainly used for modifying them.

Sketch Tools includes: Fillet, Chamfer, Offset, Convert Entities, Trim, Extend, Construction Geometry, Mirror, Stretch Entities, Move Entities, Rotate Entities, Scale Entities, Copy Entities, and Pattern (Linear Pattern, Circular Pattern).

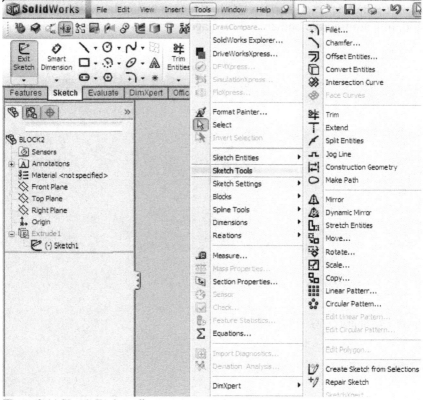

Figure 2-14 Sketch Tools toolbar

2.3.1 *Fillet*

The Fillet tool is used to create a fillet between lines in a model, in order to remove sharp edges. The sketch shown in Fig. 2-15 is filleted at a constant value of 5 mm radius between adjacent lines.

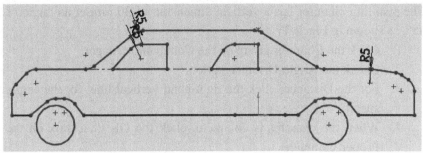

Figure 2-15 Illustrating the use of filleting

2.3.2 *Chamfer*

The Chamfer tool is used to create a chamfer in a model. Generally, a chamfer of 45° is common, but other angles could be used. Let us illustrate chamfering by considering an object, which has all of its sides either horizontal or vertical. A chamfer is needed between the vertical and horizontal lines to the right of the object, at an angle of 45° (see Fig. 2-16).

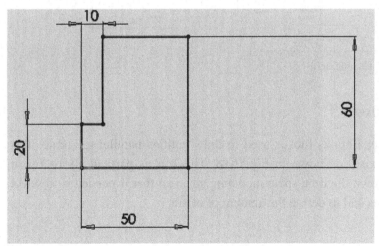

Figure 2-16 Initial geometry before applying the chamfer tool

The resulting chamfer has a vertical dimension of 30 mm at an angle of 45°, as shown in Fig. 2-17:

1. Click the Chamfer group on the CommandManager.
2. Check the Angle-Distance option.
3. For the Distance, click the right-hand vertical line; for the angle, specify 45°.
4. When the chamfer is complete, click the OK checkbox on the PropertyManager.

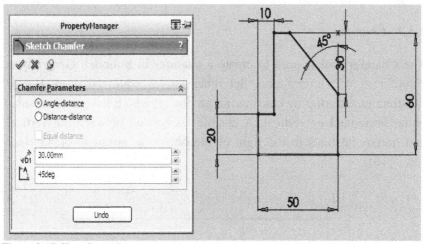

Figure 2-17 Chamfer tool

2.3.3 *Offset*

The Offset Entities tool is used to draw entities parallel to entities that already exist, as shown in Fig. 2-18. This tool is particularly useful. It can cut down the time spent modeling since all that is needed is to select the entities and to define the amount of offset.

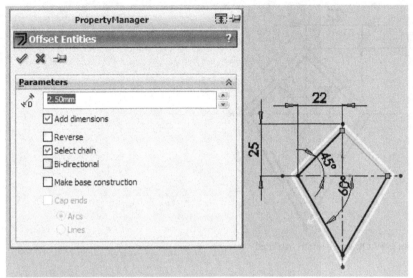

Figure 2-18 Offset Entities tools

1. Click the Offset group on the CommandManager.
2. Select and click edges to be offset.
3. Specify the offset desired (2.5 mm in this case).
4. When the offset is complete, click the OK checkbox on the PropertyManager.

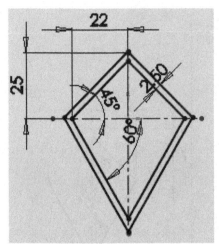

Figure 2-19 Offset geometry realized

2.3.4 *Convert Entities*

The Convert Entities tool is used to extract portions of an entity onto a plane, which are then used for further modeling, as shown in Fig. 2-20. For example, a cut-out is realized from a circle by using this tool to describe a sector. This sector could then be used to cut through the length or portion of the part being modeled (say, a cylinder).

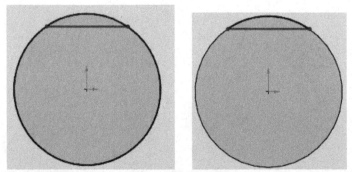

Figure 2-20 Illustrating the Convert Entities tool

2.3.5 *Trim*

The Trim tool is used to clean up lines or arcs that are not needed, as shown in Fig. 2-21.

1. Click the Trim group on the CommandManager.
2. Select the Power Trim option.
3. Hold down and drag the cursor across the entities to be trimmed.
4. When trimming is complete, click the OK checkbox on the PropertyManager (see the trimmed drawing in Fig. 2-22).

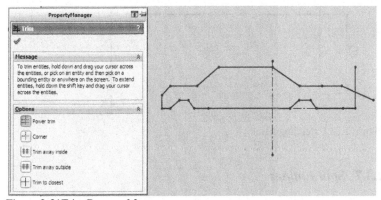

Figure 2-21 Trim PropertyManager

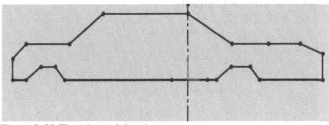

Figure 2-22 The trimmed drawing

2.3.6 *Extend*

The Extend tool is used to extend an existing line to increase its length. See Fig. 2-23. It is found in the Trim Entities group in the CommandManager.

1. Click the Extend group on the CommandManager.
2. Click the top and bottom lines to be extended.
3. When the extend task is complete, click the OK checkbox on the PropertyManager.

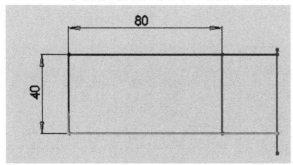

Figure 2-23 Extending a line

2.3.7 *Split Entities*

The Split Entities tool is found on the Sketching Tools menu. When it is chosen, click on the entity to trim approximately where you want the split. Fig. 2-24 shows how this tool is used.

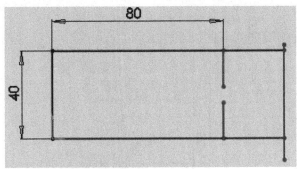

Figure 2-24 Split entities tool

2.3.8 *Construction Geometry*

The Construction Geometry option converts a solid line, circle or arc to a construction line, circle or arc, respectively.

2.3.9 *Mirror*

The Mirror tool is used to create a mirror image of an entity in a sketch. This tool is used when an axis, usually a line, exists about which a symmetrical entity or entities can be mirrored. There are two forms of the mirror tool: simple and dynamic. In the simple form, create only half of the required entities and mirror them about the axis, as shown in Fig. 2-25. In the dynamic form, as an entity is sketched, it is dynamically mirrored about the chosen axis. Mirrored entities are show in Fig. 2-26.

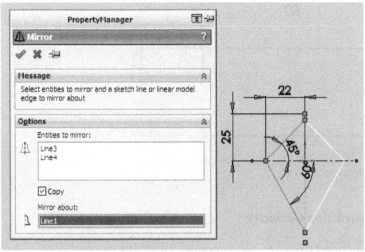

Figure 2-25 Mirror entities tool

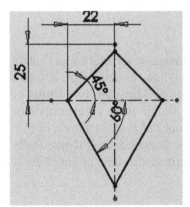

Figure 2-26 Mirrored entities

2.3.10 *Stretch Entities*

The Stretch Entities tool stretches the entities that have been chosen; the value of stretch is based on the defined datum, as shown in Fig. 2-27. In this example, the right-hand top line, the right-hand vertical line and the bottom line are stretched about the left-hand vertical line.

1. Click the Stretch group on the CommandManager.
2. Click the lines to be stretched.
3. Click the Base Defined by clicking the left-hand vertical line.
4. When the stretch task is complete, click the OK checkbox on the PropertyManager.

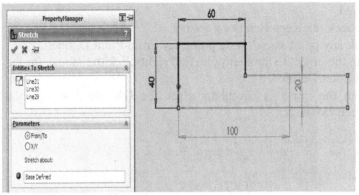

Figure 2-27 Stretch entities tool

The bottom line has been stretched from 100 mm to 125 mm while the top right line has been stretched form 40 mm to 65 mm. See Fig. 2-28 for the stretched entities.

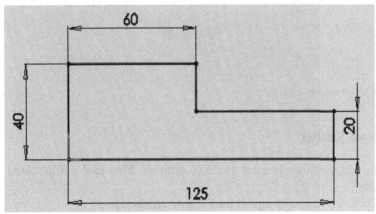

Figure 2-28 Illustrating stretched entities

2.3.11 *Move Entities*

The Move Entities tool is used to move entities. An individual item or an entire object can be moved as shown in Fig. 2-29.
1. Click the Move group on the CommandManager.
2. Click Entities to Move and place a box around the entities to be moved.
3. Uncheck the Keep Relations option.
4. Click any point as a starting point, for From Point Defined.
5. Drag the object to another point and right-click the mouse for the End Point Defined.
6. When the move is complete, click the OK checkbox on the PropertyManager.

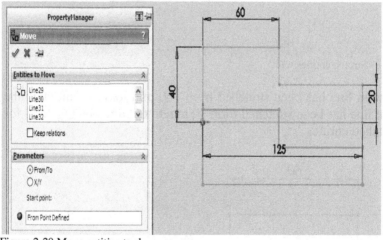

Figure 2-29 Move entities tool

2.3.12 *Rotate Entities*

The Rotate Entities tool is used to rotate entities. This tool is illustrated in Fig. 2-30.
1. Click the Move group on the CommandManager.
2. Click Entities to Rotate and place a box around the entities to be rotated.
3. Uncheck the Keep Relations option.

4. Click any point as the Center of Rotation.
5. Enter the angle of rotation.
6. When the rotation is complete, click the OK checkbox on the PropertyManager.

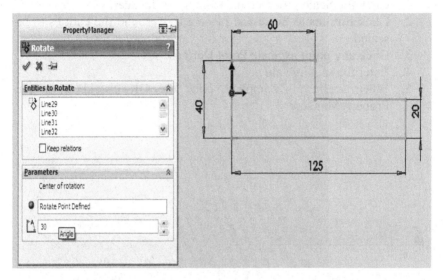

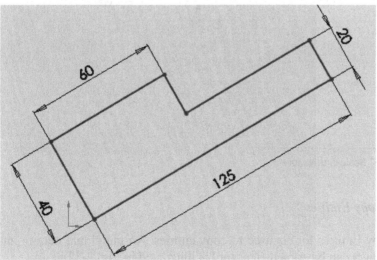

Figure 2-30 Rotate entities tool

2.3.13 *Scale Entities*

The Scale Entities tool is used to scale or change the size of entities. This tool is illustrated in Fig. 2-31.
1. Click the Scale group on the CommandManager.
2. Click Entities to Scale and place a box around the entities to be scaled.
3. Click any point as Scale Point Defined.
4. Enter the scaling value.
5. When scaling is complete, click the OK checkbox on the PropertyManager.

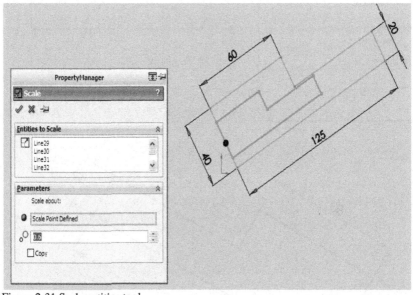

Figure 2-31 Scale entities tool

2.3.14 *Copy Entities*

The Copy Entities tool is used to copy entities. An individual item or an entire object can be copied. This tool is illustrated in Fig. 2-32.

1. Click the Copy group on the CommandManager.
2. Click Entities to Copy and place a box around the entities to be copied.
3. Uncheck the Keep Relations option.
4. Click any point as a starting point for From Point Defined.
5. Drag the object to another point and right-click the mouse for the End Point Defined.
6. When copying is complete, click the OK checkbox on the PropertyManager.

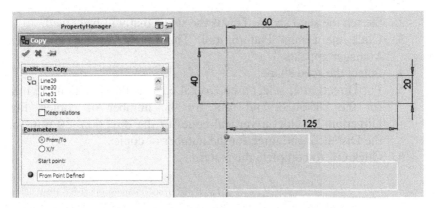

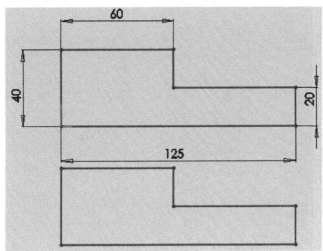

Figure 2-32 Copy entities tool

2.3.15 *Pattern*

The linear and circular pattern tools are used to create rectangular patterns by copying a given shape.

Linear pattern

For a linear pattern, directions for the patterns are linear. Let us consider the example shown in Fig. 2-33.
1. Sketch a shape.
2. Sketch the seed shape. This is the shape that will be copied.
3. Click the Linear Pattern tool. The Linear Pattern Properties Manager appears.
4. Select the seed shape.
5. For Direction 1, select a horizontal edge as the direction, 35 mm for the Distance, and four for the number of copies. For Direction 2, select a vertical edge for the direction, 30 mm for the Distance, and three for the number of copies.
6. Click OK to complete the pattern.

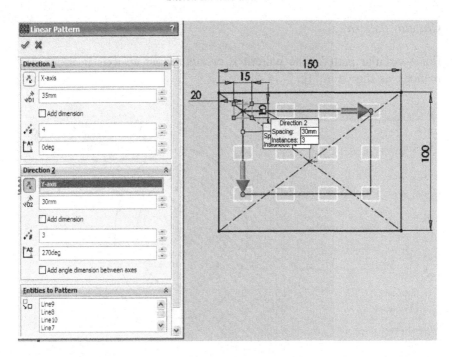

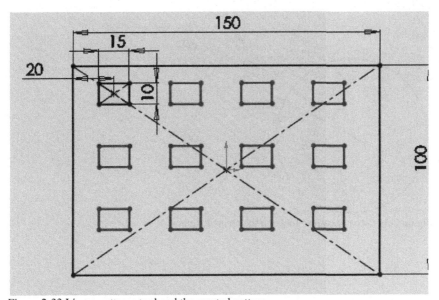

Figure 2-33 Linear patterns tool and the created pattern

Circular pattern

For a circular pattern, the pattern is created about an axis in a circular manner, as shown in Fig. 2-34.

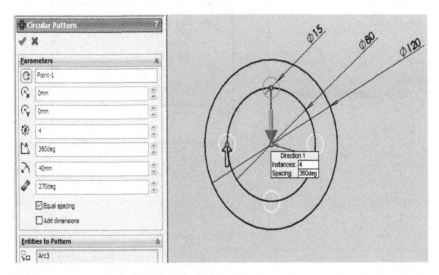

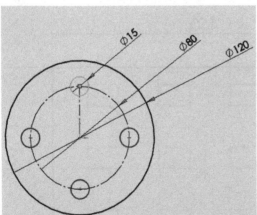

Figure 2-34 Circular pattern tool and the created pattern

Exercises

1. Use the given dimensions to redraw the shape in Fig. 2-35. Create part
models of the objects. Thickness = 15 mm

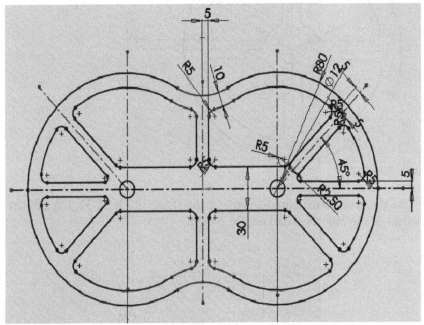

Figure 2-35

2. Use the given dimensions to redraw the shape in Fig. 2-36. Create part
models of the objects. Thickness = 25 mm

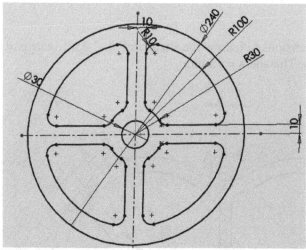

Figure 2-36

3. Use the given dimensions to redraw the shape in Fig. 2-37. Create part models of the objects. Thickness = 20 mm

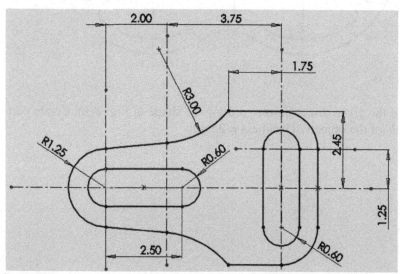

Figure 2-37

4. Use the method in the section on ellipses to create the rugby ball model shown in Fig. 2-38.

Figure 2-38 Rugby ball

5. Use the method in the section on parabolas to create the satellite model shown in Fig. 2-39.

Figure 2-39 Satellite dish

6. Use the method in the section on Convert Entities to create the cut-out cylinder model shown in Fig. 2-40.

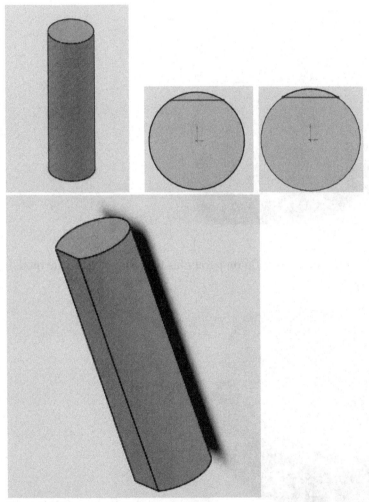

Figure 2-40 Cut-out cylinder

Chapter 3

Features

Objectives:

When you complete this chapter you will have:

- Learnt about Features tools in SolidWorks
- Learnt how to create 3D objects using Extrusion Features tools
- Learnt how to create 3D objects using Revolved Features tools
- Learnt how to create 3D objects using Lofted Features tools
- Learnt how to create 3D objects using Swept Features tools
- Learnt how to use Modification Features tools
- Learnt how to Edit Features and be able to troubleshoot models
- Learnt how to work with Reference Planes to create more complex models

3.1 Introduction

SolidWorks creates 3D objects based on Features. This chapter introduces Features tools (see Fig. 3-1) and classifies them into five categories: extrusion features (extruded boss/base, draft, dome, rib, and extruded cut), revolved features (revolved boss/base and revolved cut), lofted features (lofted boss/base and lofted cut), swept features (swept boss/base and swept cut), and modification features (hole wizard, shell, fillet, chamfer, pattern, and mirror). In the first four categories, boss/base or cut models are created. In the fifth category, tools for altering the geometry of models are introduced. How to edit features and work with reference planes to create more complex models are also introduced.

In illustrating all the Features tools presented in this chapter, it is assumed that a new document has been opened for modeling and the Document Properties have been set appropriately, and the design standard has been set as mm or inch.

Figure 3-1 SolidWorks Features tools

3.2 Extruded Boss/Base

The Extruded Boss/Base tool is used for adding height or thickness to an existing 2D sketch in order to create a 3D model. It is an *addition* feature. To use the Extruded Boss/Base tool:

1. Select the front plane, such as Sketch1 shown in Fig. 3-2.
2. Click the Features tool.
3. Click the Extruded Boss/Base tool. The Extrude Property-Manager appears.
4. Enter the extrusion height as 25mm. A real-time preview will appear (see Fig. 3-3).
5. Click OK to complete the extrusion as shown in Fig. 3-4.

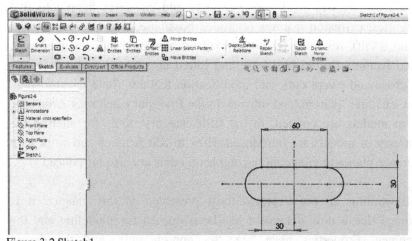

Figure 3-2 Sketch1

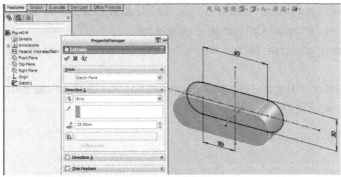

Figure 3-3 Real-time preview of extruded Sketch1

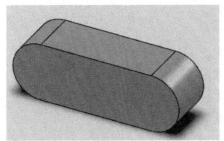

Figure 3-4 Part model obtained through use of the Extruded Boss/Base tool

3.2.1 *Draft, Dome, Rib*

Draft

The Draft tool is used to create a sloping side in a feature. Let us illustrate the tool with the example in Fig. 3-5.

1. Select the front plane and create Sketch1 (50 mm by 50 mm by 45 mm).
2. Click the Features tool.
3. Click the Draft tool. The Draft PropertyManager appears.
4. Click Manual for Type of Draft. The Neutral Plane is chosen by default.
5. Enter the Draft Angle as 30°.

6. Click the upper face (Face<1>) for the Neutral Plane. The arrow will point upwards. Correct it, if otherwise.
7. Click two adjacent faces (Face<2> and Face<3>) as Faces to Draft.
8. Click OK to complete the process of adding the draft to the part model (see Fig. 3-6).

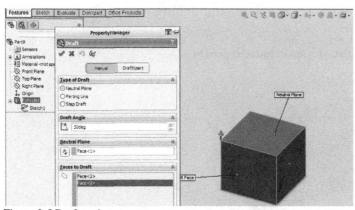

Figure 3-5 Draft tool

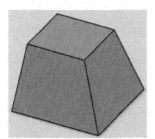

Figure 3-6 Result of adding draft features to a part model

Dome

The Dome tool is used to add a dome shape to a feature. Let us illustrate the tool with the example in Fig. 3-7.

1. Select the front plane and create Sketch1.
2. Click the Features tool.

3. Click the top face of the cylinder.
4. Click the Dome tool. The Dome PropertyManager appears.
5. Enter the dome height as 10 mm. A real-time preview will appear.
6. Click OK to complete the process of adding a dome to the part model (see Fig. 3-8).

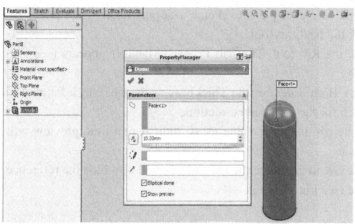

Figure 3-7 Dome tool

Figure 3-8 Result of adding a dome feature to a part model

Rib

The Rib tool is used to add a rib feature to a part. We will illustrate this tool with an L-shaped model, to which a rib will be added.

1. Select the front plane (Sketch1) for the L-shape and define a reference plane through the middle of the extruded length, as in Fig. 3-9.
2. Use the reference plane to sketch a sloping line.
3. Click the Features tool.
4. Click the Rib tool. The Rib PropertyManager appears (see Fig. 3-10).
5. Check Both Sides for the Thickness of the rib; check Parallel to Sketch for the Extrusion direction.
6. Set the rib thickness value as 10 mm. A real-time preview will appear.
7. Click OK to complete the rib and this will also hide the reference plane (see Fig. 3-11).

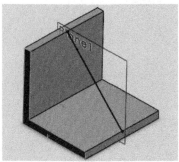

Figure 3-9 Sketching a sloping line on Plane1 of the L-shape to define a rib

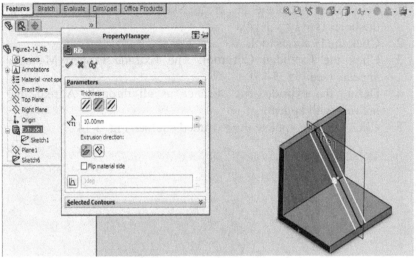

Figure 3-10 Rib PropertyManager

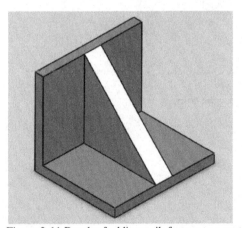

Figure 3-11 Result of adding a rib feature to a part model

3.2.2 *Extruded Cut*

The Extruded Cut tool is used to remove a portion of an existing 2D sketch or part during 3D modeling. It is a *subtraction* feature. Let us illustrate this tool by making two holes, each 10 mm in diameter, located 30 mm from each side of the center of the 3D model shown in Figs 3-12 to 3-14. To use the Extruded Cut tool:

1. Select the front plane and create Sketch2 (two circles) to define the holes (see Fig. 3-12).
2. Click the Features tool.
3. Click the Extruded Cut tool. The Extrude PropertyManager appears (see Fig. 3-13).
4. Define the extruded-cut distance as all through. A real-time preview will appear.
5. Click OK to complete the extruded cut (see Fig. 3-14).

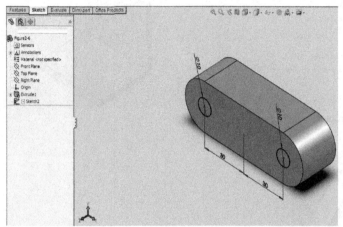

Figure 3-12 Creating circles, Sketch2 defines the holes

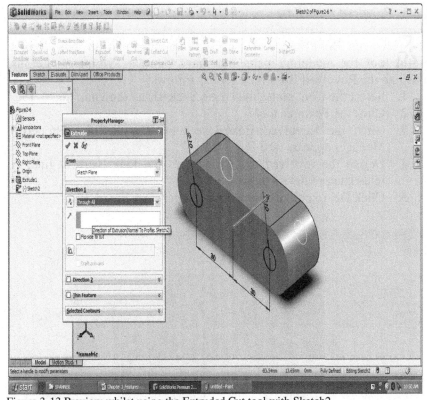

Figure 3-13 Preview whilst using the Extruded Cut tool with Sketch2

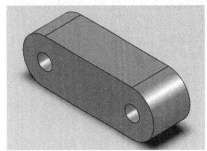

Figure 3-14 Resulting part model obtained through using the Extruded Cut tool

3.2.3 *Revolved Boss/Base*

The Revolved Boss/Base tool rotates a contour about an axis. It is a useful tool for modeling parts that have circular symmetry. Let us illustrate the Revolved Boss/Base tool as follows:

1. Select the front plane and create Sketch1, as shown in Fig. 3-15.
2. Click the Features tool.
3. Click the Revolved Boss/Base tool. The Revolve Property-Manager appears (see Fig. 3-16).
4. Define the revolved axis (Line1) as the long, vertical line. A real-time preview will appear.
5. Click OK to complete the revolved part (see Fig. 3-17).

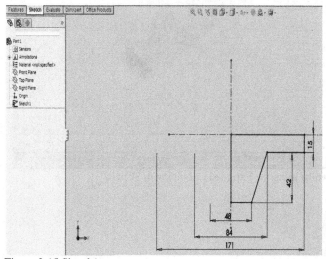

Figure 3-15 Sketch1

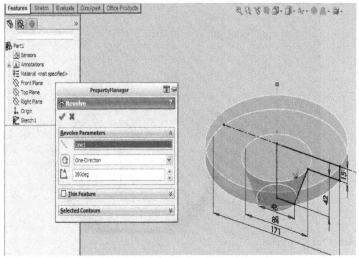

Figure 3-16 Revolve PropertyManager Preview

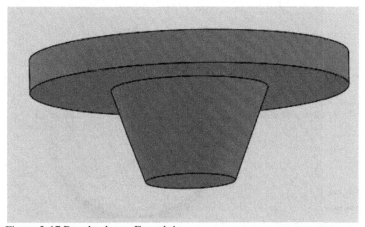

Figure 3-17 Revolved part, Extrude1

Creating four holes on the revolved part

The steps involved in creating four holes are as follows:
1. Sketch four holes, each 15 mm diameter, as shown in Fig. 3-18.
2. Extrude-cut each hole, using Through All, as shown in Fig. 3-19. The final revolved part with holes is shown in Fig. 3-20.

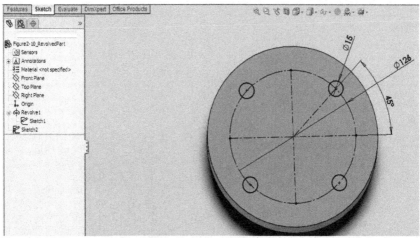

Figure 3-18 Sketches for four holes

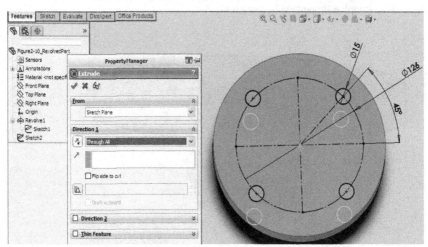

Figure 3-19 Extrude-cut each hole

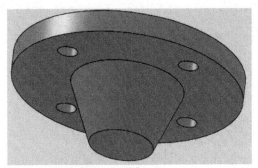

Figure 3-20 Final revolved feature

3.2.4 *Revolved Cut*

The Revolved Cut tool is for removing revolved sections out of 3D objects. It is a *subtraction* feature. Let us illustrate the tool by cutting a hole from the center of the 3D model created in the previous section. To use the Revolved Cut tool:

1. Select the front plane and create Sketch3 (a rectangle) to define the hole (see Fig. 3-21).
2. Click the Features tool.
3. Click the Revolved Cut tool. The Revolve PropertyManager appears (see Fig. 3-22).
4. Define the revolved axis (Line1) as the vertical dimension line. A real-time preview will appear.
5. Click OK to complete the revolved cut (see Fig. 3-23).

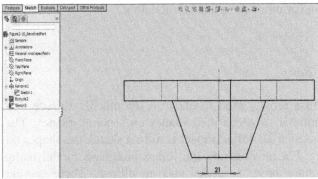

Figure 3-21 Sketch for revolve-cut

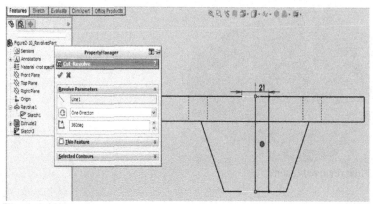

Figure 3-22 Revolve-cut

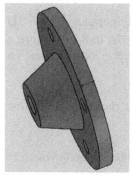

Figure 3-23 Final 3D part created using the Revolved Cut tool

Although defining a circle and cutting through does the same thing, this exercise has shown the steps required in a creating a revolved cut feature, which may be useful for more complicated modeling.

3.2.5 *Lofted Boss/Base*

The Lofted Boss/Base tool is used to create a smooth 3D surface that passes through a number of predefined shapes on separate planes. The prerequisite for creating a 3D lofted model is to first sketch the shapes on the different planes. Let us illustrate the lofted boss/base model using three shapes: a square base, a circle, and an ellipse. There are no restrictions on the shapes that could be used. Let us illustrate the lofted boss/base concept as follows:

1. Select the top plane and create Sketch1 (a square, 90 mm by 90 mm) as shown in Fig. 3-24.
2. Exit sketch mode.
3. Click the Features tool and click the Reference tool. Select the Plane option. The plane box appears; select the top plane.
4. Set the distance between the existing top plane and a new reference plane as 50 mm. Click the OK checkbox. A new plane, Plane1, appears as shown in Fig. 3-25.
5. Create Sketch2 (a circle of diameter 60 mm) as shown in Fig. 3-26.
6. Exit sketch mode.
7. Click the Features tool and click the Reference tool. Select the Plane option. The plane box appears; select the top plane.
8. Set the distance between Plane1 and a new reference plane as 50 mm. Click the OK checkbox. A third plane, Plane2, appears as shown in Fig. 3-27.
9. Create Sketch3 (an ellipse of major diameter 80 mm, and minor diameter 65 mm) as shown in Fig. 3-27.
10. Exit sketch mode. All sketches appear as shown in Fig. 3-27.
11. Click the Lofted Boss/Base tool. The Loft PropertyManager appears.
12. Right-click the Profiles box.
13. Click the square, then the circle, then the ellipse. A real-time preview will appear, as shown in Fig. 3-28.
14. Click OK to complete the lofted part. Hide the planes, see Fig. 3-29.

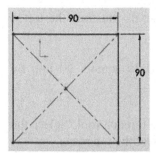

Figure 3-24 Sketch1

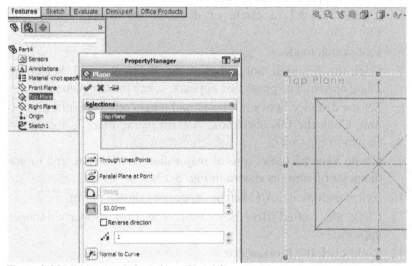

Figure 3-25 Middle sketch for lofting (Sketch2)

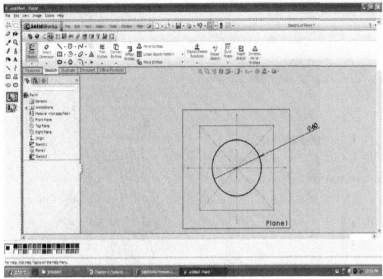

Figure 3-26 Top sketch for lofting (Sketch3)

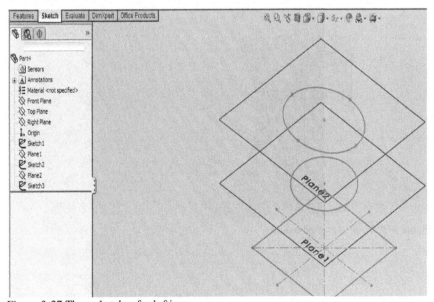

Figure 3-27 Three sketches for lofting

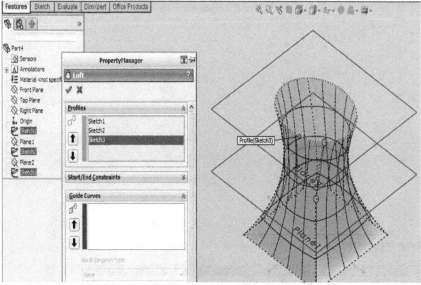

Figure 3-28 Preview of loft based on Sketch1, Sketch2, and Sketch3

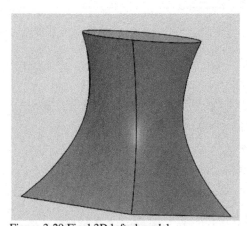

Figure 3-29 Final 3D lofted model

3.2.6 *Lofted Cut*

The Lofted Cut tool is used to cut a shape between a number of planes, each of which contains a defined shape or geometry. The prerequisite for creating a 3D lofted model is to first sketch the shapes on the different planes. Let us illustrate the lofted cut model using a square base, a circle, and an ellipse, similar to the previous model obtained using a lofted boss/base, but this time offset into the model (see Fig. 3-30).

1. Select the top plane. Click the Features tool and click the Reference tool. Select the Plane option. The plane box appears; select the top plane. Set the distance between the existing top plane and a new reference plane for 5 mm further away from the 3D model. Click the OK checkbox. A new plane, Plane3, appears.
2. Create Sketch4 (a square, 83 mm by 83 mm).
3. Exit sketch mode.
4. Click the Features tool and click the Reference tool. Select the Plane option. The plane box appears; select the top plane.
5. Set the distance between the existing Plane1 and a new reference plane for 5 mm below. Click the OK checkbox. A third plane, Plane4, appears.
6. Create Sketch5 (a circle of diameter, 57 mm).
7. Exit sketch mode.
8. Click the Features tool and click the Reference tool. Select the Plane option. The plane box appears; select the top plane.
9. Set the distance between Plane2 and a new reference plane for 5 mm above. Click the OK checkbox. A fourth plane, Plane5, appears.
10. Create Sketch3 (an ellipse of major diameter, 67 mm, and minor diameter, 60 mm).
11. Exit sketch mode.
12. Click the Lofted Cut tool. The Loft PropertyManager appears.
13. Right-click the Profiles box.
14. Click the new square, then the new circle, then the new ellipse. A real-time preview will appear.
15. Click OK to complete the lofted cut part, which is now hollow. Hide the planes.

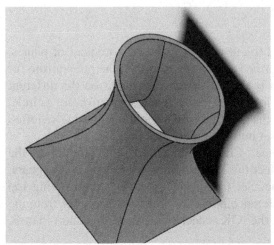

Figure 3-30 Result of using the Lofted Cut tool

3.2.7 *Swept Boss/Base*

The Swept Boss/Base tool is used to sweep a profile through a path (arc, spline, etc). As with lofting, the shapes have to be created first. The prerequisite for creating a 3D swept model is to sketch the shapes on different perpendicular planes. In the illustration presented here, a hexagon is the *profile* while a spline is the *path*, as shown in Fig. 3-31.

1. Select the top plane. Create Sketch1 (a hexagon with sides of 0.5 in).
2. Exit sketch mode.
3. Select the front plane. Create Sketch2 (a spline with one end at the center of the hexagon).
4. Exit sketch mode.
5. Click the Swept Boss/Base tool. The Sweep PropertyManager appears (see Fig. 3-32).
6. Right-click the Profile and Path box.
7. Click the hexagon as the profile and the spline as the path. A real-time preview will appear.
8. Click OK to complete the swept part (see Fig. 3-33). Hide the planes.

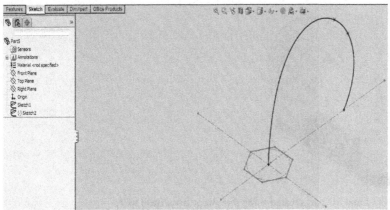

Figure 3-31 Profile (hexagon) and path (spline)

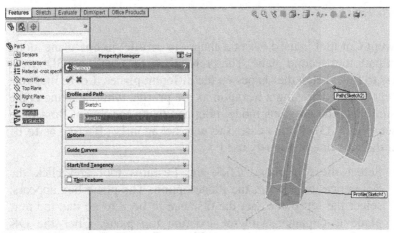

Figure 3-32 Preview and Swept PropertyManager

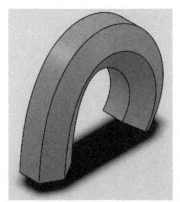

Figure 3-33 Final 3D part created using the Swept Boss/Base tool

3.2.8 *Swept Cut*

The Swept Cut tool is used to cut a shape (or profile) in one plane along a given path to a second plane. The prerequisite for creating a 3D swept model is to first sketch the shapes on the different planes. Let us illustrate the swept cut model using a hexagon as the profile, while a spline is the path, similar to the previous model obtained using the Swept Boss/Base tool (see Fig. 3-33). In this case, the new hexagon is offset so as to be smaller than the one previously used.

1. Select the top plane. Click the Features tool and click the Reference tool. Select the Plane option. The plane box appears; select the top plane. Set the distance to be zero so that the new plane is coplanar with the existing top plane. Click the OK checkbox. A new plane, Plane3, appears.
2. Create Sketch2 (Sketch5) (a hexagon with sides offset inward by 0.05 in), as in Fig. 3-34.
3. Exit sketch mode.
4. Click the Swept Cut tool. The Swept PropertyManager appears (see Fig. 3-35).
5. Right-click the Profile and Path box.

6. Click the hexagon as the profile and the spline as the path. A real-time preview will appear.
7. Click OK to complete the swept cut part, which is now hollow. Hide the planes (see Fig. 3-36).

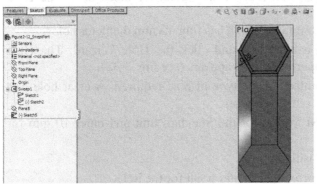

Figure 3-34 Sketch for cutting

Figure 3-35 Preview and Swept PropertyManager

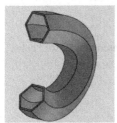

Figure 3-36 Final 3D part created using the Swept Cut tool

3.2.9 *Hole Wizard*

The Hole Wizard is used to add hole(s) to an existing 3D model, as shown in Fig. 3-37. Let us use the revolved boss/base model as an example; we want to add four 10 mm diameter holes using the Hole Wizard.

1. Click the front face, which has four 15 mm diameter holes.
2. Click the Features tool and click the Hole Wizard. The Hole Wizard PropertyManager appears (see Fig. 3-38).
3. Click the button for the type of hole required. A clear hole was selected for our illustration.
4. Select ANSI Metric for the Standard unit and enter 10 mm for the Size.
5. Click the Position button.
6. Click an approximate center point for the hole.
7. Click OK. A dialog will appear (see Fig. 3-39). Use the Smart Dimension tool to locate the center of the hole (see Fig. 3-40).
8. Click OK.
9. The hole will be added to the model (see Fig. 3-41).

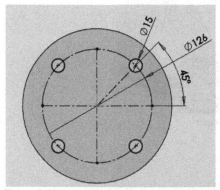

Figure 3-37 An existing model, to which the Hole Wizard is to be used to add extra holes

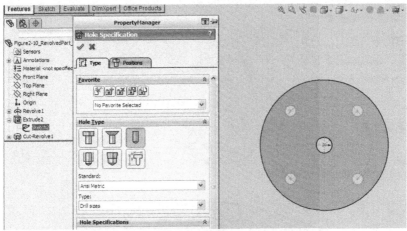

Figure 3-38 Hole Wizard used to create an additional, central hole

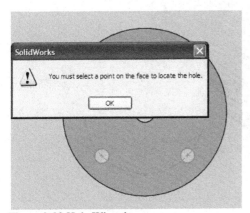

Figure 3-39 Hole Wizard message

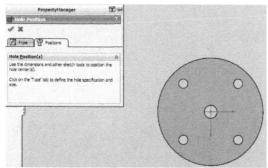

Figure 3-40 Position for central hole

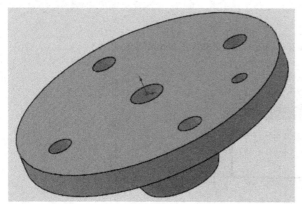

Figure 3-41 Hole created using the Hole Wizard

3.2.10 *Shell*

The Shell tool is used to create a hollow shape within an existing 3D model, transforming it into a hollow object having uniform thickness. Let us use the lofted model of Fig. 3-29 to illustrate the Shell tool by applying it to the square base. The thickness of the shell will be 5 mm.

1. Select the square base (see Fig. 3-42).
2. Click the Shell tool. The Shell PropertyManager appears (see Fig. 3-43).
3. Apply the thickness of 2.5 mm.
4. Click OK to complete the shell part, which is now hollow as shown in Fig. 3-44.

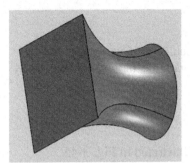

Figure 3-42 Bottom is chosen for shelling

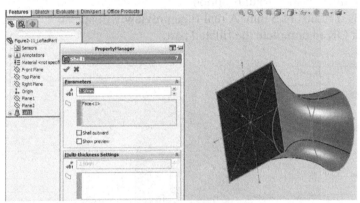

Figure 3-43 Shell thickness defined

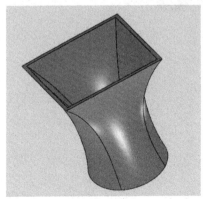

Figure 3-44 Final 3D model created using the Shell tool

3.2.11 *Fillet Tool*

A fillet is a rounded corner. There are usually four ways of defining a fillet: having a constant radius, having a variable radius, by face or by full round, as shown in Figs 3-45 to 3-48.

Defining a constant radius fillet

Defining a fillet using a constant radius is illustrated in Fig. 3-45.
1. Click the Fillet tool. The Fillet PropertyManager appears.
2. Click the Constant Radius option.
3. Apply a radius of 5 mm. A real-time preview will appear.
4. Click OK to complete the filleted part.

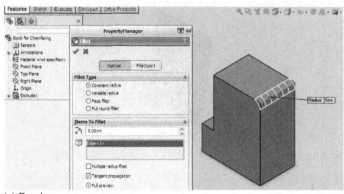

(a) Preview

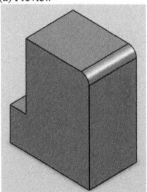

(b) Filleted part
Figure 3-45 Fillet tool for creating constant radius fillets

Defining a variable radius fillet

Defining a fillet using a variable radius is illustrated in Fig. 3-46.

1. Click the Fillet tool. The Fillet PropertyManager appears.
2. Click the Variable Radius option.
3. Click on an edge to apply the fillet. Two boxes appear on the screen, one at each end of the edge to accept the values.
4. Enter 2.5 mm at the word Unassigned in the first box.
5. Enter 5 mm at the word Unassigned in the second box. A real-time preview will appear.
6. Click OK to complete the filleted part.

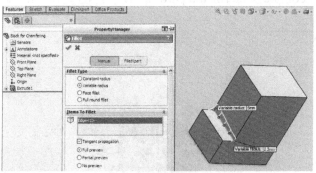

(a) Preview

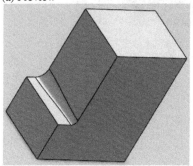

(b) Filleted part

Figure 3-46 Fillet tool for creating variable radius fillets

Defining a face fillet

Defining a fillet using a face is illustrated in Fig. 3-47.

1. Click the Fillet tool. The Fillet PropertyManager appears.
2. Click the Face fillet option.
   Two boxes appear in Items to Fillet; these are used to define the two faces of the fillet.
3. Apply a radius of 5 mm.
4. Define Face1 by clicking on a face as shown.
5. Click the second box in Items to Fillet and define Face2 by clicking on the face as shown. A real-time preview will appear.
6. Click OK to complete the filleted part.

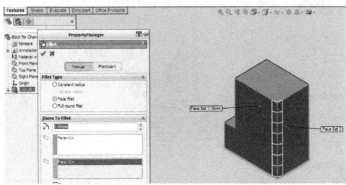

(a) Preview

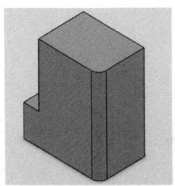

(b) Filleted part

Figure 3-47 Fillet tool for creating face fillets

Defining a full round fillet

Defining a full round fillet is illustrated in Fig. 3-48.

1. Click the Full round fillet tool. The Fillet PropertyManager appears.
2. Click the Full round fillet option.
   Three boxes appear in Items to Fillet; these are used to define the three faces of the fillet.
3. Define Face1 by clicking on a face as shown.
4. Click the second box in Items to Fillet and define Face2 by clicking on the face as shown.
5. Click the third box in Items to Fillet and define Face3 by clicking on the face as shown. A real-time preview will appear.
6. Click OK to complete the filleted part.

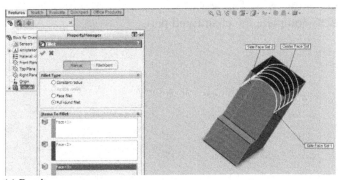

(a) Preview

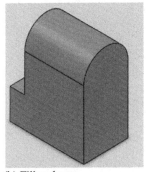

(b) Filleted part
Figure 3-48 Fillet tool for creating full round fillets

3.2.12 *Chamfer Tool*

A chamfer is a slanted surface added to the corner of a part. Chamfers are usually manufactured at $45°$ but any other angle may be used. There are usually three ways of defining a chamfer: with a distance and an angle (e.g. $2.5×45°$), with two distances (e.g. $2.5×2.5$) or at a vertex as shown in Figs 3-49 to 3-51.

Defining a chamfer using an angle and distance

Defining a chamfer using an angle and distance is illustrated in Fig. 3-49.
1. Click the Chamfer tool. The Chamfer PropertyManager appears.
2. Click the option Angle distance.
3. Enter a distance of 2.5 mm and accept the default angle, $45°$. A real-time preview will appear.
4. Click OK to complete the chamfered part.

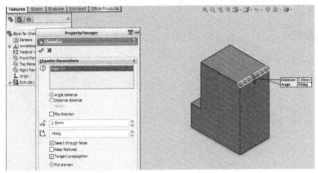

(a) A real-time preview

(b) Chamfered part

Figure 3-49 Chamfer tool for creating chamfers defined using an angle and a distance

Defining a chamfer using two distances

Defining a chamfer using two distances is illustrated in Fig. 3-50.

1. Click the Chamfer tool. The Chamfer PropertyManager appears.
2. Click the option Distance distance.
3. Enter 2.5 mm for both distances. A real-time preview will appear.
4. Click OK to complete the chamfered part.

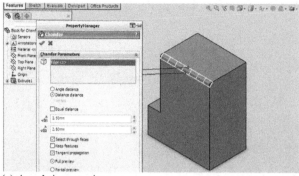

(a) A real-time preview

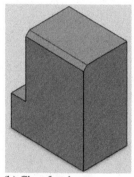

(b) Chamfered part

Figure 3-50 Chamfer tool for creating chamfers defined using two distances

Defining a chamfer at a vertex

Defining a chamfer at a vertex is illustrated in Fig. 3-51.

1. Click the Chamfer tool. The Chamfer PropertyManager appears.
2. Click the Vertex tool. Three distance boxes will appear.
3. Define the three distances. Although 2.5 mm is used, any other value could be used.
4. A real-time preview will appear.
5. Click OK to complete the chamfered part.

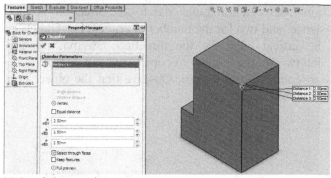

(a) A real-time preview

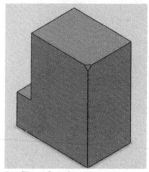

(b) Chamfered part
Figure 3-51 Chamfer tool for creating a chamfer at a vertex

3.2.13 *Linear Pattern*

The Linear Pattern tool is used to create rectangular patterns by copying a given shape. Let us consider an example.
1. Sketch a rectangle, 100 mm by 50 mm and extrude it 5 mm.
2. Click the bottom and sketch one small rectangle, 5 mm by 5 mm; extrude it 50 mm.
3. Click the Linear Pattern tool. The Linear Pattern Property-Manager appears.
4. Select the small rectangle from the FeatureManager as the Feature to Pattern.

5. For Direction 1, select a horizontal edge for the direction, 70 mm for the Distance, and two for the number of copies. For Direction 2, select a vertical edge for the direction, 36 mm for the Distance, and two for the number of copies (see Fig. 3-52).
6. Click OK to complete the pattern. Observe that the object is a table, as shown in Fig. 3-53.

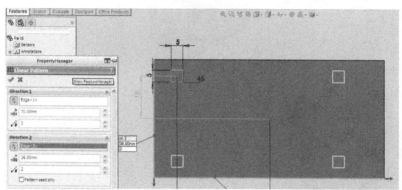

Figure 3-52 Defining distances and number of copies in Direction 1 and Direction 2

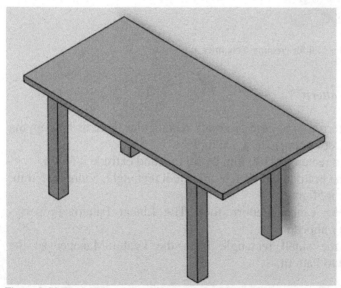

Figure 3-53 Final 3D model created using the Linear Pattern tool

3.2.14 *Circular Pattern*

The Circular Pattern tool is used to create a circular pattern on a 3D model about an origin. The 10mm diameter hole created using the Hole Wizard, see Fig. 3-14, will now be duplicated four times in a circular manner.

1. Click the Circular Pattern tool. The Circular Pattern PropertyManager appears (see Fig. 3-54).
2. Select the 10mm diameter hole created using the Hole Wizard from the FeatureManager for the Feature to Pattern.
3. Click View > Temporary Axis to activate the temporary axes.
4. Select the axis through the origin of the model (Axis<1>), about which to pattern.
5. Define the number of copies.
6. Click OK to complete the pattern (see Fig. 3-55).

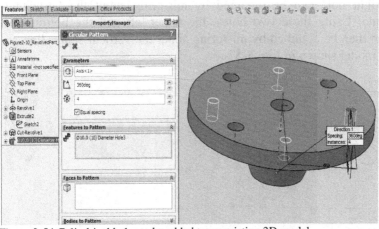

Figure 3-54 Cylindrical holes to be added to an existing 3D model

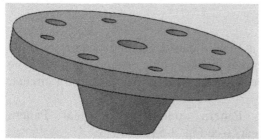

Figure 3-55 Final 3D model created using the Circular Pattern tool

3.2.15 *Mirror*

The Mirror tool is used to create mirror images of features. The axes for mirroring have to be defined. For example, for the object shown in Fig. 3-56, if only the hole in the upper-right side is initially defined, the bottom-right hole can be obtained by mirroring about the horizontal center line. Using these two holes, the upper-left and the bottom-left holes can then be obtained by mirroring about the vertical center line, to give the model shown in Fig. 3-57.

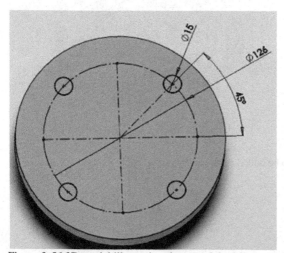

Figure 3-56 3D model illustrating the use of the Mirror tool

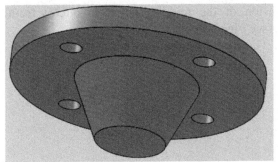

Figure 3-57 Final 3D model with hole features obtained using the Mirror tool

3.2.16 *Reference Planes*

The Reference Planes tool is a very powerful tool for modeling complex objects. Reference planes can be defined relative to the standard planes (top, right, and left) or relative to faces on a model. Reference planes were used for the lofting examples in Sections 3.2.5 and 3.2.6. They were also used in other models but hidden after completion of the modeling exercise. Reference planes are very useful and should be generously used. But there is a catch! Rotate the model sufficiently to ensure that your reference planes are exactly where you want them to be, otherwise you will sketch in the wrong place.

3.2.17 *Editing Features*

SolidWorks Edit Features tools are extremely powerful. The key to understanding the Edit Features tools is understanding the FeatureManager design tree. The design tree gives the history of how the components of a part were modeled, as shown in Fig. 3-58. Since features make up a part, and sketches make up a feature, we can easily edit a part by simply editing its features or its sketches. It is that easy, but you need to try it out for different parts and master the process for yourself. Let us edit the sketch of the lofted model, shown in Fig. 3-29, and see how this will affect the final shape.
1. Right-click Sketch2 in the FeatureManager.
2. Click Edit Sketch.

3. Change the dimension of the middle sketch (a circle) from 60 mm to 100 mm (see Fig. 3-59).

4. Click Exit Sketch. The model bulges outward as shown in Fig. 3-60!

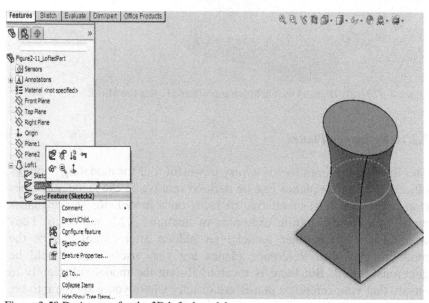

Figure 3-58 Design tree for the 3D lofted model

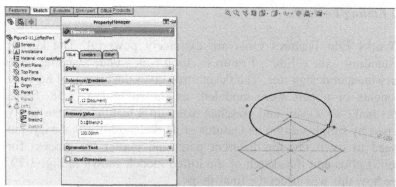

Figure 3-59 Editing the middle sketch (Sketch2)

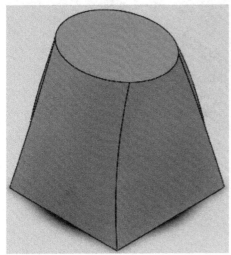

Figure 3-60 Edited model

SolidWorks Edit Features tools are powerful for editing models via the design tree. By merely changing one dimension of the middle sketch (the circle) in the middle plane, a completely different shape has been created. The designer can change a number of dimensions to form a suite of different designs. This is a useful tool for industrial designers.

3.2.18 *Tutorials*

Tutorial 3.1: Simple Part 3-1

In this tutorial we will create the model shown in Fig. 3-61. The sketch of the model is shown in Fig. 3-62. We will create the sketch and fully dimension it. The model will be extruded to a depth of 1.25 in. We will determine the center of mass for the model if it is made of 1060 alloy. Note the origin of the model.

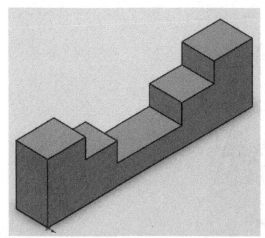

Figure 3-61 Model for Tutorial 3.1

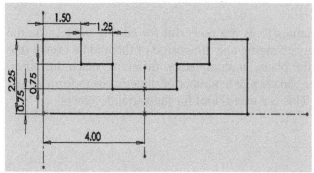

Figure 3-62 Sketch for Tutorial 3.1

1. Open a New SolidWorks part document
2. Set the document properties for the model, with decimal places equal to two.
3. Create Sketch1, as shown in Fig. 3-62.
4. Click the Features tool.
5. Click the Extruded Boss/Base tool. The Extrude Property-Manager appears.
6. Enter an extrusion depth of 1.25 in. A real-time preview will appear.

7. Click OK to complete the extrusion, Extrude1, as shown in Fig. 3-61.
8. Assign the 1060 alloy to the part modeled.
9. Calculate the mass properties of the part (see Fig. 3-63).

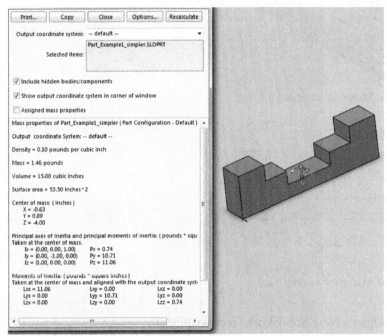

Figure 3-63 Mass properties of the part for Tutorial 1

Tutorial 3.2: Simple Part 3-2

We will create the model shown in Fig. 3-64. The sketch of the model is shown in Fig. 3-65. We will create the sketch and fully dimension it. The model will be extruded a depth of 1.25 in. We will determine the center of mass for the model if it is made of 1060 alloy. Note the origin of the model.

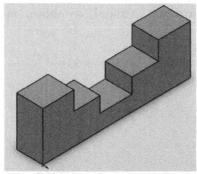

Figure 3-64 Model for Tutorial 3.2

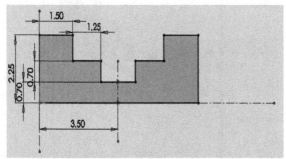

Figure 3-65 Sketch for Tutorial 3.2

1. Open a New SolidWorks part document.
2. Set the document properties for the model, with decimal places equal to 2.
3. Create Sketch1 as shown in Fig. 3-65.
4. Click the Features tool.
5. Click the Extruded Boss/Base tool. The Extrude Property-Manager appears.
6. Enter an extrusion depth of 1.25 in. A real-time preview will appear.
7. Click OK to complete the extrusion, Extrude1, as shown in Fig. 3-64.
8. Assign the 1060 alloy to the part modeled.
9. Calculate the mass properties of the part (see Fig. 3-66).

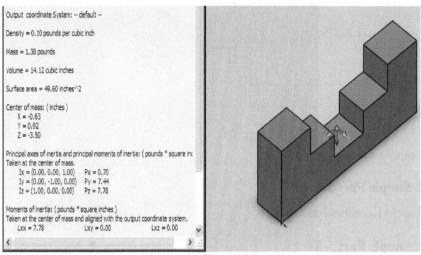

Figure 3-66 Mass properties of the part for Tutorial 3.2

Exercises

1. Simple Part 3-3

Create the model shown in Fig. 3-67. The sketch of the model is shown in Fig. 3-68. Create the sketch and fully dimension it. Extrude the model to a depth of 1.25 in. Determine the center of mass for the model if it is made of 1060 alloy. Note the origin of the model.

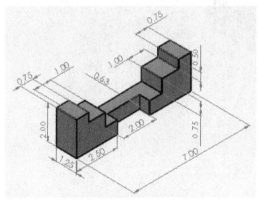

Figure 3-67 Model for Exercise 1

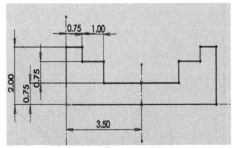

Figure 3-68 Sketch for Exercise 1

2. Simple Part 3-4

Repeat Exercise 2 using brass.

3. Simple Part 3-5

Create the model shown in Fig. 3-69. The sketch of the model is shown in Fig. 3-70. Create the sketch and fully dimension it. Extrude the model to a depth of 1.25 in. Determine the center of mass for the model if it is made of 1060 alloy. Note the origin of the model.

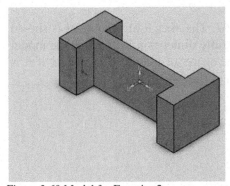

Figure 3-69 Model for Exercise 3

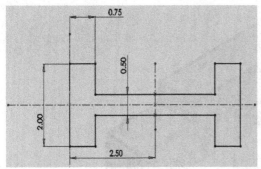

Figure 3-70 Sketch for Exercise 3

4. Simple Part 3-6

Repeat Exercise 3, but setting the height of sketch to 3.00 in (see Fig. 3-71) and using brass.

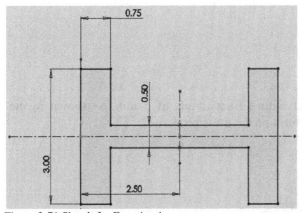

Figure 3-71 Sketch for Exercise 4

5. Simple Part 3-7

Create the model shown in Fig. 3-72. Create the sketch and fully dimension it. Determine the center of mass for the model if it is made of brass. Note the origin of the model.

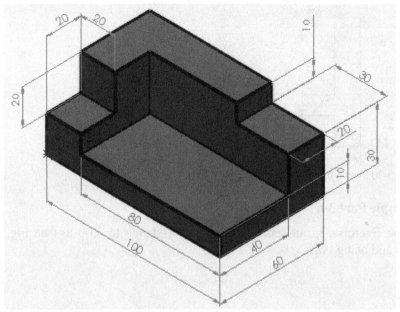

Figure 3-72 Model for Exercise 5

6. Simple Part 3-8

Repeat Exercise 5, but with a fillet radius of 5 mm, as shown in the model in Fig. 3-73. Determine the mass properties.

Figure 3-73 Part for Exercise 6 with a fillet radius of 5 mm

7. Simple Part 3-9

Create the model shown in Fig. 3-74, which also shows the sketch. Create the sketch and fully dimension it. Note the symmetry and origin of the model. Determine the center of mass for the model if it made of 1060 alloy.

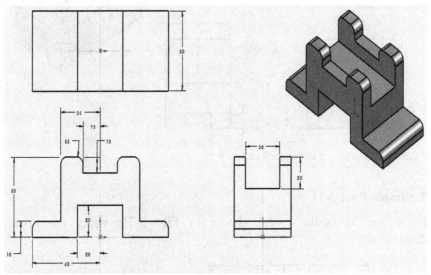

Figure 3-74 Model and sketch for Exercise 7

8. Simple Part 3-10

Create the model shown in Fig. 3-75, which also shows the sketch. Create the sketch and fully dimension it. Note the symmetry and origin of the model. Determine the center of mass for the model if it is made of 1060 alloy.

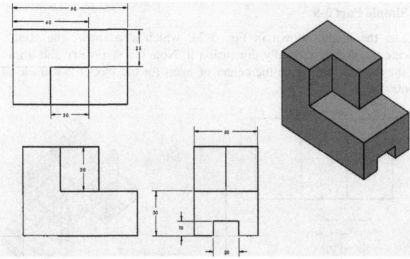

Figure 3-75 Model and sketch for Exercise 8

9. Simple Part 3-11

Draw the 3D solid model shown in Fig. 3-76, using the given dimensions.

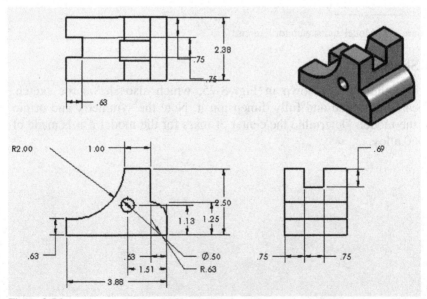

Figure 3-76

10. Simple Part 3-12

Draw the 3D solid model shown in Fig. 3-77, using the given dimensions.

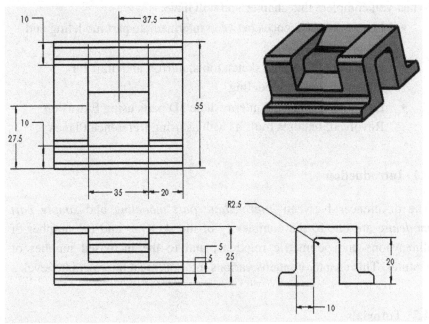

Figure 3-77

Chapter 4

Intermediate Part Modeling

Objectives:

When you complete this chapter you will have:

- Learnt the differences between intermediate part modeling and simple part modeling
- Learnt how to utilize sketch tools, mirror, and draft for intermediate part modeling
- Learnt how to model intermediate 3D parts using Extrusion, Revolved, features tools as well as using Reference Planes

4.1 Introduction

The differences between *intermediate part modeling* and *simple part modeling* are due to the complexity of the sketches and the number of dimensions and geometric relations, and to the increased number of features. This chapter contains various tutorials at an intermediate level.

4.2 Tutorials

Tutorial 4.1a: Widget

We will build the widget shown in Fig. 4-1. The lines and the arcs are tangential to each other. Determine the center of mass for this part, if it is made from the 1060 alloy.

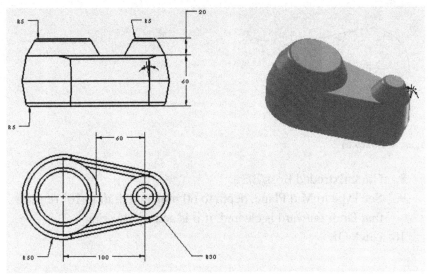

Figure 4-1 Widget

1. Open a New SolidWorks part document.
2. Sketch a horizontal center line through the origin.
3. Sketch a slanting line on one side of the center line.
4. Click the Mirror tool.
5. Mirror the slanting line about the horizontal center line.
6. The center of the 30 mm radius arc is 60 mm to the right of the origin; and the center of the 50 mm radius arc is 40 mm to the left of the origin.
7. Add a Tangent Relation between each of the lines and the two arcs. Trim any dangling edges. See Fig. 4-2 for the sketch so far, Sketch1.

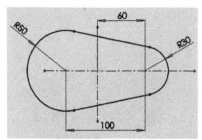

Figure 4-2 Sketch1

8. Click Extruded Boss/Base.
9. Set Type to Mid Plane, depth to 60 mm, and draft to 10° (ensure that Draft outward is cleared, if it is active) (see Fig. 4-3).
10. Click OK.

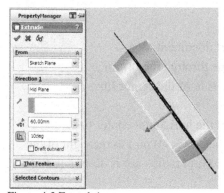

Figure 4-3 Extrude1

Add Bosses

11. Sketch two circles using the two existing centers of the existing arcs (highlight the arcs to *awake* the centers).
12. Add the Co-radial Relation between the larger circle and the larger arc, see Fig. 4-4.
13. Add the Co-radial Relation between the smaller circle and the smaller arc.
14. Click Extruded Boss/Base.

15. Set Type to Blind, depth to 20 mm, and draft to 30° (ensure that Draft outward is cleared, if it is active), see Fig. 4-5.

16. Click OK (the final modeled part is shown in Fig. 4-6).

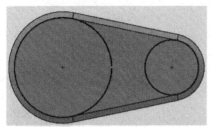

Figure 4-4 Sketches for second extrusion

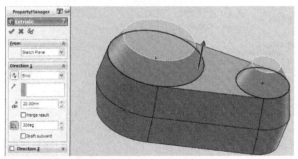

Figure 4-5 Preview

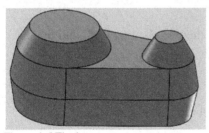

Figure 4-6 Final part

Center of mass

Right-click Material > Edit Material in FeatureManager. Choose 1060 alloy and Apply. Click Evaluate > Mass Properties (see mass calculation output in Fig. 4-7).

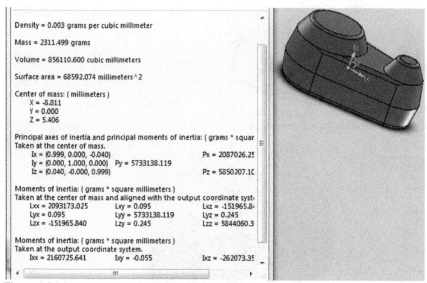

Density = 0.003 grams per cubic millimeter

Mass = 2311.499 grams

Volume = 856110.600 cubic millimeters

Surface area = 68592.074 millimeters^2

Center of mass: (millimeters)
  X = -8.811
  Y = 0.000
  Z = 5.406

Principal axes of inertia and principal moments of inertia: (grams * squar
Taken at the center of mass.
  Ix = (0.999, 0.000, -0.040)   Px = 2087026.25
  Iy = (0.000, 1.000, 0.000)   Py = 5733138.119
  Iz = (0.040, -0.000, 0.999)   Pz = 5850207.1C

Moments of inertia: (grams * square millimeters)
Taken at the center of mass and aligned with the output coordinate syst
  Lxx = 2093173.025   Lxy = 0.095   Lxz = -151965.8
  Lyx = 0.095   Lyy = 5733138.119   Lyz = 0.245
  Lzx = -151965.840   Lzy = 0.245   Lzz = 5844060.3

Moments of inertia: (grams * square millimeters)
Taken at the output coordinate system.
  Ixx = 2160725.641   Ixy = -0.055   Ixz = -262073.3$

Figure 4-7 Mass properties of part

Tutorial 4.1b: Widget

Repeat Tutorial 4.1a where the ends of the two slanting lines and the centers of the arcs are collinear (i.e. they have vertical relations).

Modeling

As Tutorial 4.1a with the appropriate relations applied.

Center of mass

Right-click Material > Edit Material in FeatureManager. Choose 1060 alloy and Apply. Click Evaluate > Mass Properties (see mass calculation output in Fig. 4-8).

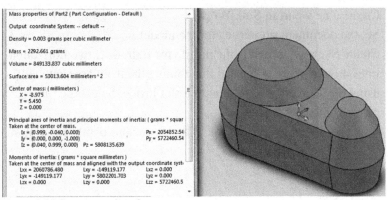

Figure 4-8 Mass properties of part

Tutorial 4.2

We will build the part shown in Fig. 4-9. A sketch of the part is shown in Fig. 4-10. We will determine the center of mass for this part, as if it is made from cast alloy steel, which has a density of 0.0073 g mm^{-3}.

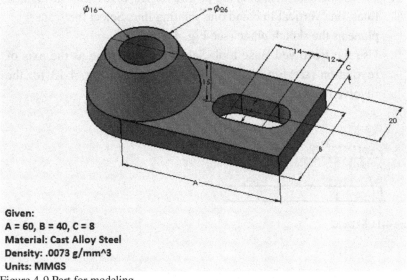

Given:
A = 60, B = 40, C = 8
Material: Cast Alloy Steel
Density: .0073 g/mm^3
Units: MMGS
Figure 4-9 Part for modeling

1. Create a new part in SolidWorks.
2. Set the document properties for the model.
3. Create Sketch1, which is the profile for Extrude1: two horizontal lines 40 mm apart, a vertical line joining them, an arc sketched using the Tangent Arc tool, with a slot inside. Select the top plane as the sketch plane.
4. Use the Extrude Base tool. The extrusion depth is 8 mm. Extrude1 is the base feature.

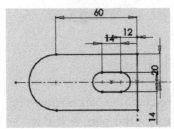

Figure 4-10 Sketch1

5. Create Sketch2, which is the profile for Revolve1: two horizontal lines, one vertical line and one slanting line. Select the front plane as the sketch plane (see Fig. 4-11).
6. Use the Revolved Base tool. Select the centerline as the axis of revolution (see Fig. 4-12 for the preview and Fig. 4-13 for the revolved part).

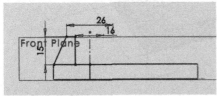

Figure 4-11 Sketch2

114

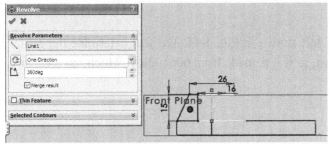

Figure 4-12 Preview of the revolved part

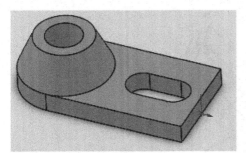

Figure 4-13 Final revolved part

The mass properties are shown in Fig. 4-14.

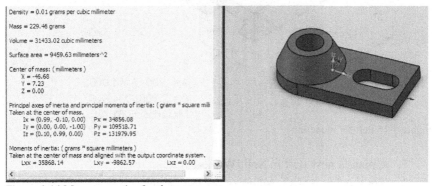

Figure 4-14 Mass properties for the part

Tutorial 4.3a

We will build the part shown in Fig. 4-15. We will determine the center of mass for this part, as if it is made from 6061 alloy, which has a density of 0.097 lb in^{-3}.

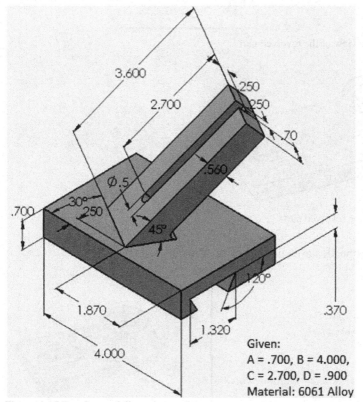

Figure 4-15 Part for modeling

1. Create a new part in SolidWorks.
2. Set the document properties for the model.
3. Create Sketch1, 4.00 in by 2.70 in. Select the top plane as the sketch plane (see Fig. 4-16).
4. Use the Extrude Base tool. The extrusion depth is 0.7 in. Extrude1 is the base feature (see Fig. 4-17).

116

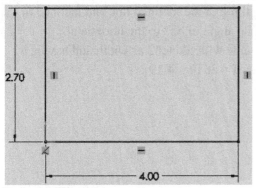

Figure 4-16 Sketch1

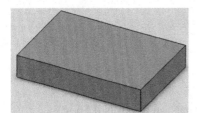

Figure 4-17 Extrude1

5. Create Sketch2, a line 1.40 in long, inclined at an angle of 30° to the horizontal, located 1.87 in from a right-hand vertical edge and 0.25 in from the orthogonal, adjacent edge. Select the top of Extrude1 as the sketch plane (see Fig. 4-18).

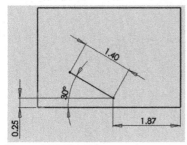

Figure 4-18 Sketch2

6. Create Plane1, passing through the inclined line and inclined to the top of Extrude1, at an angle of 45° to the horizontal.
7. Create Sketch3, a rectangle with Sketch2 as width and having a length of 3.6 in, on Plane1 (see Fig. 4-19).

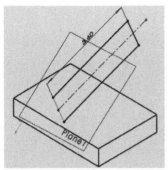

Figure 4-19 Sketch3

8. Use the Extrude Boss tool. The extrusion depth is 0.56 in. Extrude2 is located on Plane1 (see Fig. 4-20).

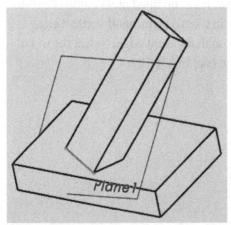

Figure 4-20 Extrude2

9. Create Sketch4, a profile for the first extruded cut feature. Select the top face of the inclined Extrude2 as the Sketch Plane (see Fig. 4-21).

10. Use the Extruded Cut tool. Select Blind as the End Condition, the depth is 0.25 in.

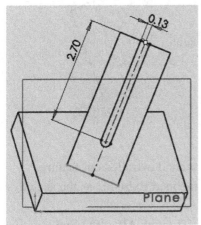

Figure 4-21 Sketch4

11. Use the Fillet tool at the back edge of the inclined Extrude2, the radius is 0.12 in (see Fig. 4-22).

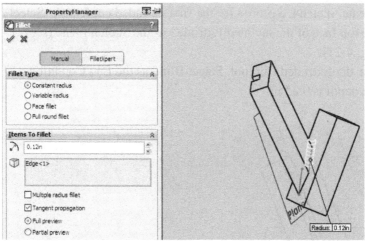

Figure 4-22 Fillet additions

12. Create Sketch5, a profile for the second extruded cut feature. Select the right-hand face of the inclined Extrude2 as the sketch plane (see Fig. 4-23).

13. Use the Extruded Cut tool. Select Through All as the End Condition.

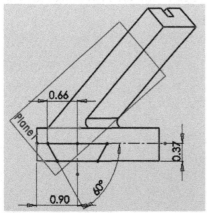

Figure 4-23 Sketch5

The completed part is shown in Fig. 4-24, while the material properties are shown in Fig. 4-25.

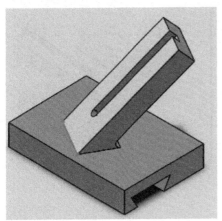

Figure 4-24 Final part

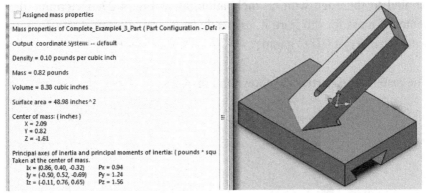

Assigned mass properties

Mass properties of Complete_Example4_3_Part (Part Configuration - Def $\triangle$

Output coordinate system: -- default

Density = 0.10 pounds per cubic inch

Mass = 0.82 pounds

Volume = 8.38 cubic inches

Surface area = 48.98 inches^2

Center of mass: (inches)
    X = 2.09
    Y = 0.82
    Z = -1.61

Principal axes of inertia and principal moments of inertia: (pounds * squ
Taken at the center of mass.
    Ix = (0.86, 0.40, -0.32)   Px = 0.94
    Iy = (-0.50, 0.52, -0.69)   Py = 1.24
    Iz = (-0.11, 0.76, 0.65)   Pz = 1.56

Figure 4-25 Material properties of part

Tutorial 4.3b

Modify the units of measure for the part in Tutorial 4.3a, from IPS to MMGS. Modify the material from cast alloy steel, 6061 alloy, to ABS. Modify the Plane1 angle from 45° to 30°. Determine the center of mass for this part.

The material properties for this part are shown in Fig. 4-26.

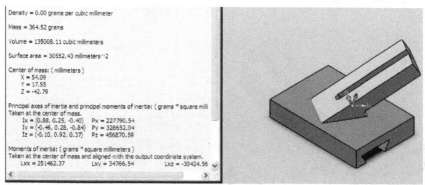

Figure 4-26 Material properties

Tutorial 4.4

Build the part for the previous tutorial. Modify Sketch2, so that it is parallel to the edge with no inclination, as in Fig. 4-27. Determine the center of mass for this part, if it is made from cast alloy steel, 6061 alloy, with a density of 0.097 g mm^{-3}.

The material properties are shown in Fig. 4-28.

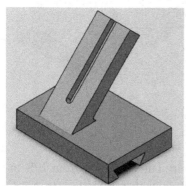

Figure 4-27 Final modeled part

Mass properties of Complete_Modified_Example4_3_Part (Part Configura ▲

Output coordinate System: -- default --

Density = 0.10 pounds per cubic inch

Mass = 0.82 pounds

Volume = 8.38 cubic inches

Surface area = 48.98 inches^2

Center of mass: (inches)
    X = 1.83
    Y = 0.82
    Z = -1.57

Figure 4-28 Material properties for modified part

Tutorial 4.5a

We will build the part shown in Fig. 4-29. A = 76 mm, B = 127 mm, material is 2014 alloy, density = 0.0028 g mm^{-3}. All fillets are equal, 6 mm. Determine the center of mass for this part.

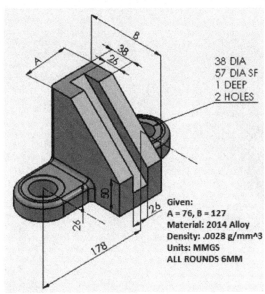

Figure 4-29 Part for modeling

1. Create a new part in SolidWorks.
2. Set the document properties for the model.
3. Create Sketch1, a profile with two vertical lines, a horizontal line, and an inclined line. Select the front plane as the sketch plane (see Fig. 4-30).
4. Use the Extrude Base tool. The extrusion depth is 76 mm. Extrude1 is the base feature (see Fig. 4-31).

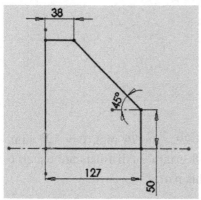

Figure 4-30 Sketch1

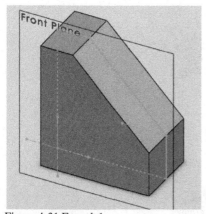

Figure 4-31 Extrude1

5. Create Sketch2, a square, 26 mm on each side. Select the flat top of Extrude1 as the sketch plane (see Fig. 4-32).
6. Use the Extruded Cut tool. Select All Through as the End Condition. Select an inclined edge as the Direction of Extrusion (see Fig. 4-32).

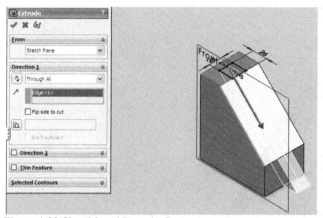

Figure 4-32 Sketch2 and Extrude-Cut1

7. Create Sketch3, a profile with two vertical lines, an arc at their end points, using the Tangent Arc tool, and a circle. Select the bottom of Extrude1 as the sketch plane (see Fig. 4-33).
8. Use the Extrude Boss tool. Select Blind as the End Condition. The extrusion depth is 26 mm. Extrude2 is the boss feature (see Fig. 4-34).

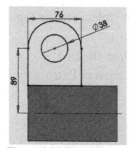

Figure 4-33 Sketch3

Figure 4-34 Boss feature, Extrude2

9. Create a mirror copy of the Extrude2 feature. The front plane is used for mirroring (see Fig. 4-35).

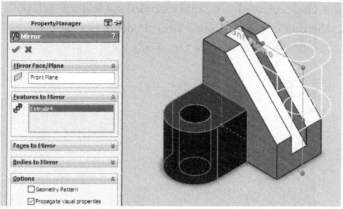

Figure 4-35 Mirror of Extrude2

10. Add fillets to the Extrude2 feature. The fillet radius is 6 mm (see Fig. 4-36).

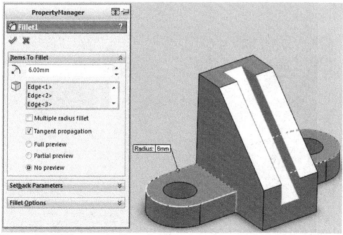

Figure 4-36 Fillets on the Extrude2 features

11. Create Sketch4, a circle of diameter 57 mm. Select the top of Extrude2 as the sketch plane (see Fig. 4-37).

12. Using this circle, create the second extruded cut feature. Select Blind as the End Condition. The extrusion depth is 1 mm. Extrude2 is the boss feature (see Fig. 4-38).

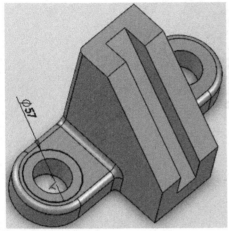

Figure 4-37 Sketch4

Figure 4-38 Second extruded cut

13. Create a mirror copy of the seconded extruded cut feature. The front plane is used for mirroring (see Fig. 4-39).

The material properties are shown in Fig. 4-40.

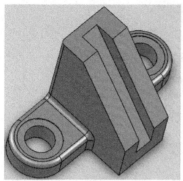

Figure 4-39 Final part

Output coordinate System: -- default --

Density = 0.0028 grams per cubic millimeter

Mass = 3432.5988 grams

Volume = 1225928.1397 cubic millimeters

Surface area = 101032.4420 millimeters^2

Center of mass: (millimeters)
    X = 49.2493
    Y = 46.8736
    Z = -0.0009

Figure 4-40 Material properties

Tutorial 4.5b

Modify the fillet radius of Extrude2 from 6 mm to 8 mm. Modify the material from the 2014 alloy to the 6061 alloy. Modify the angle of the inclined line in Sketch1 from 45° to 30°. Determine the center of mass for this part.

The part is shown in Fig. 4-41. The material properties are shown in Fig. 4-42.

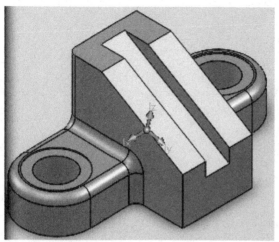

Figure 4-41 Modified part

Output coordinate System: -- default --

Density = 0.0027 grams per cubic millimeter

Mass = 3017.4542 grams

Volume = 1117575.6474 cubic millimeters

Surface area = 92968.1047 millimeters^2

Center of mass: (millimeters)
  X = 49.8041
  Y = 34.2544
  Z = 0.0255

Figure 4-42 Material properties

Tutorial 4.6a

We will build the part shown in Fig. 4-43. A = 52 mm, B = 58 mm, the material is 6061 alloy, density = 0.0027 g mm^{-3}. All fillets are equal, 4 mm. Determine the center of mass for this part.

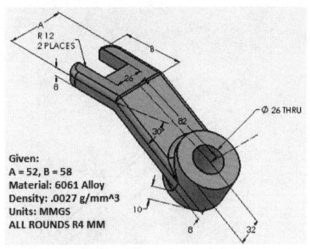

Figure 4-43 Part for modeling

1. Create a new part in SolidWorks.
2. Set the document properties for the model.
3. Create Sketch1, a profile with a horizontal line, and an inclined line at 30° to the horizontal. Select the front plane as the sketch plane (see Fig. 4-44).
4. Create the Extrude-Thin1 feature. Apply Symmetry. Select Mid Plane for the End Condition in Direction 1. The extrusion depth is 52 mm and the thickness is 12 mm. Extrude1 is the base feature (see Fig. 4-45).

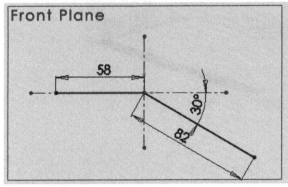

Figure 4-44 Sketch1

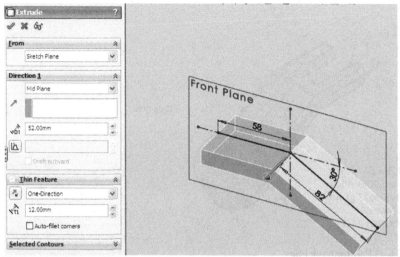

Figure 4-45 Extrude-Thin1

5. Create Plane1. Select the mid-point and the top face of the inclined edge of Extrude1. Plane1 is located halfway between the top and bottom of faces of the inclined component of Edge1 (see Fig. 4-46).

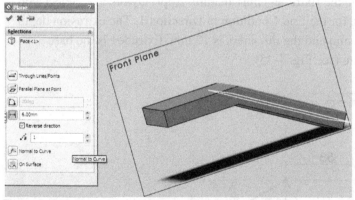

Figure 4-46 Plane1

6. Create Sketch2. Select Plane1 as the sketch plane. Sketch a circle, diameter 52 mm. Note the way Extrude1 is oriented when Plane1 is selected (see Fig. 4-47).

7. Use the Extrude Boss tool. Apply Symmetry. Select Mid Plane for the End Condition in Direction 1. The extrusion depth is 52 mm and the thickness is 32 mm (see Fig. 4-48).

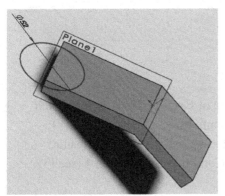

Figure 4-47 Sketch2

8. Create Sketch3. Select the top of Extrude2, the circular face, as the sketch plane. Sketch a circle, diameter 26 mm (see Fig. 4-48).

9. Using this circle, create the first extruded cut feature. Select the Through All End Condition in Direction 1 (see Fig. 4-49).

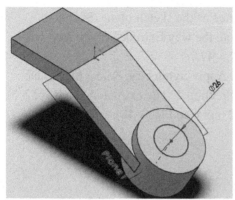

Figure 4-48 Extrude2

10. Create Sketch4. Select the top of Extrude-Thin1 as the sketch plane. Sketch a rectangle, 26 mm by 32 mm (see Fig. 4-49).
11. Using this rectangle, create the second extruded cut feature. Select the Through All End Condition in Direction 1 (see Fig. 4-50).

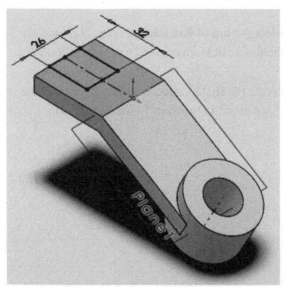

Figure 4-49 First extruded cut and Sketch4

12. Create three sets of fillets (see Figs 4-50 to 4-52).

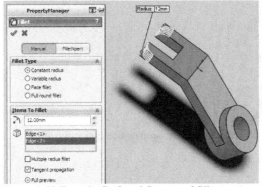

Figure 4-50 Extrude-Cut2 and first set of fillets

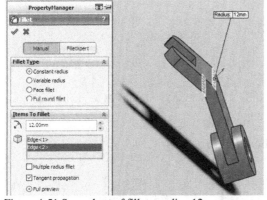

Figure 4-51 Second set of fillets, radius 12 mm

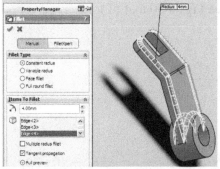

Figure 4-52 Third set of fillets, radius 4 mm; the tangent propagation option was used

Figure 4-53 Final part

The material properties are shown in Fig. 4-54.

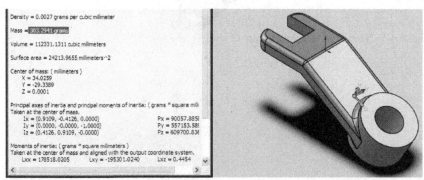

Figure 4-54 Material properties

Tutorial 4.6b

Reduce the radius for the first set of fillets from 12 mm to 10 mm and the radius for the second and third sets from 4 mm to 2 mm. Modify the angle of the inclined line in Sketch1 (see Fig. 4-44) from 30° to 45°. Reduce the extrusion depth of Extrude-Thin1 from 52 mm to 38 mm. Change the material from 6061 alloy to ABS. Recalculate the location of the center of mass with reference to the origin.

Table 4-1 Changes to part for Tutorial 4.6b

	Modify From	Modify To
Radius for first set of fillets	12 mm	10 mm
Radius for second and third sets of fillets	4 mm	2 mm
Sketch1 angle	30°	45°
Extrude-Thin1 depth	52 mm	38 mm
Material	6061 alloy	ABS

The final part and its material properties are shown in Fig. 4-55.

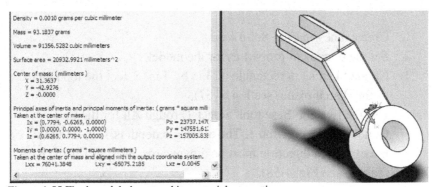

Figure 4-55 Final modeled part and its material properties

Tutorial 4.7a

We will build the part shown in Fig. 4-56. A = 3 in, B = 1 in, the material is 6061 alloy, density = 0.097 lb in^{-3}. All fillets are equal, 4 mm. Determine the center of mass for this part.

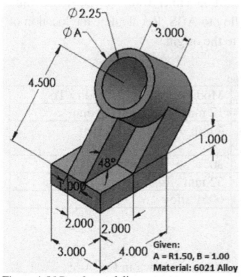

Figure 4-56 Part for modeling

1. Create a new part in SolidWorks.
2. Set the document properties for the model.
3. Create Sketch1, a rectangle of 3 in by 4 in. Select the top plane as the sketch plane (see Fig. 4-57).
4. Use the Extrude Base tool. Select Through All for the End Condition in Direction 1. The extrusion depth is 1 in. Caution: the extrusion direction must be downwards. Extrude1 is the base feature (see Fig. 4-58).

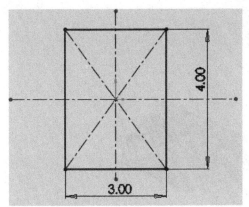

Figure 4-57 Sketch1

Figure 4-58 Extrude Base

5. Create Plane1. Select the mid-points of the sides of the top face of Extrude1, and draw a construction line between them. Plane1 is defined using this construction line. It is created at an angle of 48° to the top face of Extrude1 (see Fig. 4-59).

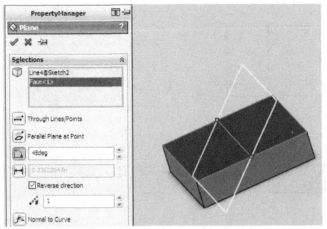

Figure 4-59 Plane1

6. Create Sketch2. It is made up of a horizontal line on Extrude1, two lines on Plane1 and an arc using the Tangent Arc tool. Select Plane1 as the sketch plane (see Fig. 4-60).

7. Use the Extrude Boss tool to create Extrude2. Select Up To Vertex for the End Condition in Direction 1. Select the top-right vertex of Extrude1. Extrude2 is the base boss. Fig. 4-61 shows a preview of Extrude2 and Fig. 4-62 shows the part once Extrude2 has been created.

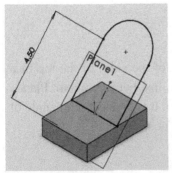

Figure 4-60 Plane1

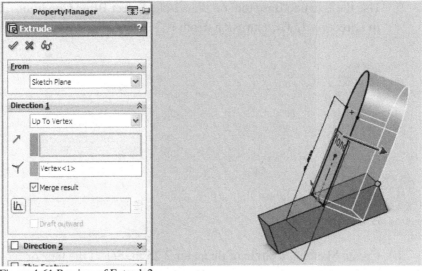

Figure 4-61 Preview of Extrude2

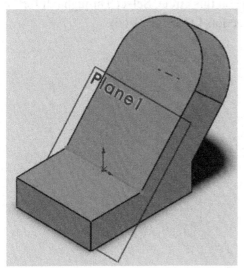

Figure 4-62 Part with Extrude2

8. Create Sketch3, a circle. Select the back angle of Extrude2 as the
    sketch plane (see Fig. 4-63).

141

9. Use the Extrude Base tool on this circle. Select Blind Condition in Direction 1. The extrusion depth is 3 in (see Fig. 4-64).

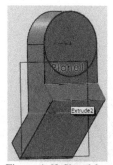

Figure 4-63 Sketch3

10. Create Sketch4, a circle of diameter 2.25 in. Select the face or back of Extrude2 (cylinder) as the sketch plane (see Fig. 4-65).
11. Use the Extruded Cut tool on this circle. Select Through All for the End Condition in Direction 1 (see Fig. 4-66).

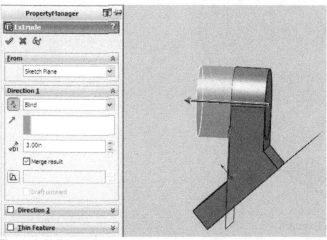

Figure 4-64 Extrude3

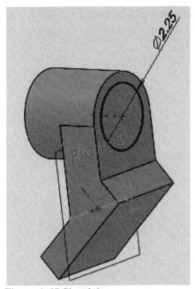

Figure 4-65 Sketch4

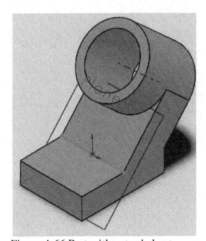

Figure 4-66 Part with extruded cut

12. Create Sketch5, a profile for the rib feature. Select the right-hand plane as the sketch plane. Apply a parallel relation (see Fig. 4-67).

13. Create the rib feature, Rib1. Its thickness is 1 in (see Fig. 4-68 for a preview of the rib feature).

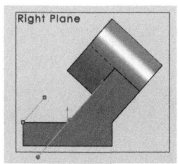

Figure 4-67 Sketch5

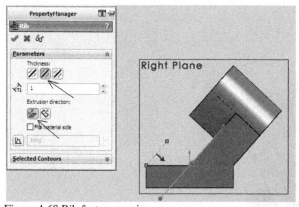

Figure 4-68 Rib feature preview

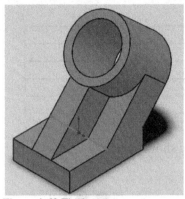

Figure 4-69 Final part

14. The material is 6061 alloy.
15. Calculate the mass properties. Fig. 4-70 shows the results.

Output coordinate System: -- default --

Density = 0.098 pounds per cubic inch

Mass = 2.989 pounds

Volume = 30.645 cubic inches

Surface area = 100.964 inches^2

Center of mass: (inches)
   X = 0.000
   Y = 0.729
   Z = -0.861

Figure 4-70 Mass properties

Tutorial 4.7b

Increase the thickness of Rib1 from 1.00 in to 1.25 in and the depth of Extrude2 from 3.00 in to 3.25 in. Modify the angle that Plane1 makes with Extrude1 from 48° to 30°. Change the material from 6061 alloy to copper. See Table 4-2 for these modifications. Recalculate the location of the center of mass with reference to the origin.

Table 4-2 Changes to part for Tutorial 4.7b

	Modify From	Modify To
Thickness of Rib1	1.00 in	1.25 in
Depth of Extrude2	3.00 in	3.25 in
Plane1 angle	48°	30°
Material	6061 alloy	Copper

The final part and its material properties are shown in Fig. 4-71.

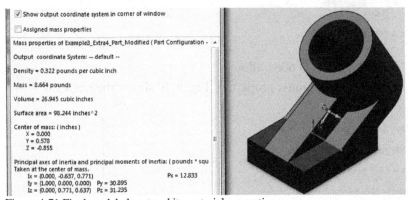

Figure 4-71 Final modeled part and its material properties

Tutorial 4.8

We will design a part in SolidWorks given the following information and using Fig. 4-72:

> Units: MMGS
> Material: 5MM, 6061 alloy
> Density: 0.0027 g mm^{-3}
> Part origin: halfway between the left-hand side and right-hand
> side of the model.
> Note: All holes are 6 mm

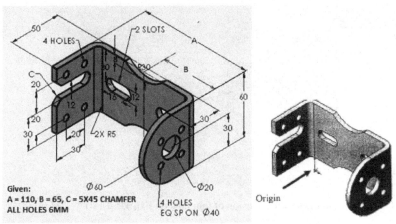

Figure 4-72 Part description

1. Create Sketch1, which is the base sketch. Select the top plane. Place the origin at the intersection of the middle of the back side and the tip of the left side of the sketch (see Fig. 4-73).

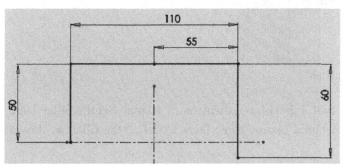

Figure 4-73 Sketch1

2. Create the Extrude-Thin1 feature. Symmetry is applied, with Mid Plane as the End Condition in Direction 1. The depth is 60 mm and the thickness is 5 mm. Check the Auto-fillet box and define the radius as 5 mm. See Fig. 4-74 for a preview of the extrusion and Fig. 4-75 for the extruded part, Extrude-Thin1.

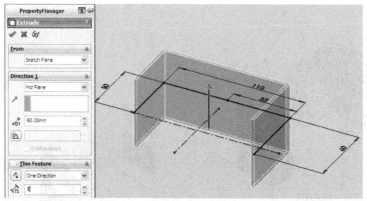

Figure 4-74 PropertyManager for extrusion of thin feature, Extrude-Thin1

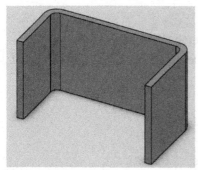

Figure 4-75 Extrude-Thin1

3. Use the Fillet tool. Fillet1 has a diameter of 40 mm. Set the Fillet Type to Full round fillet, and choose three faces to define the fillet, as shown in Fig. 4-76.

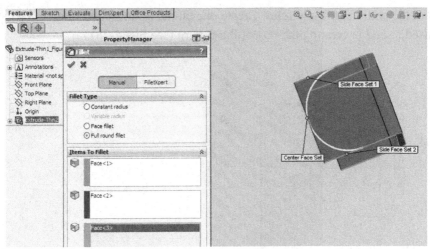

Figure 4-76 Fillet feature created

3. Create Sketch2. Select the right-hand face as the sketch plane. Sketch a circle with diameter of 20 mm (see Fig. 4-77).

4. Create the first extruded cut, the feature Extrude-Cut1. Select Up To Next as the default End Condition.

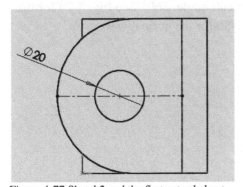

Figure 4-77 Sketch2 and the first extruded cut

5. Create Sketch3. Select the right-hand face as the sketch plane. Sketch a construction circle with a diameter of 40 mm, which will be used to lay

out the pattern of four circles. Create one circle, of diameter 6 mm, to be used a seed in creating the rest of the pattern (see Fig. 4-78).

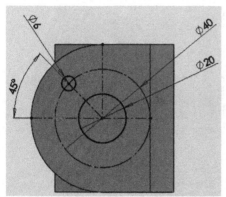

Figure 4-78 Sketch3, a seed circle for the circular pattern

6. Using the seed circle, create the second extruded cut, feature Extrude-Cut2. Select Up To Next as the default End Condition.

7. Use the Circular Pattern tool. Select Up To Next as the default End Condition. Activate the temporary axes. Use the axis at the center of the 20 mm diameter circle as the axis of rotation (Axis<1>). Set the number of instances to 4 and accept the default angle, 360°. See Fig. 4-79 for the circular pattern preview, and Fig. 4-80 for the normal view.

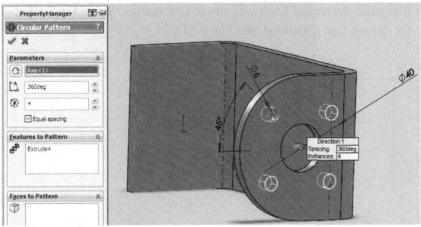

Figure 4-79 Circular pattern preview

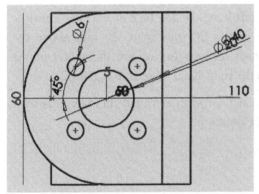

Figure 4-80 Normal view of circular pattern

8. Create Sketch4. Select the left-hand outside face of Extrude-Thin1 as the sketch plane. Sketch two lines, 12 mm apart. The lower line should pass through the origin of the part. Both lines are horizontal and 30 mm in length. Apply a Tangent Arc to the left-hand ends of these lines. Close the shape by sketching a vertical line to the right. This creates a slot (see Fig. 4-81).

9. Using the slot, create the third extruded cut feature. Select Up To Next as the default End Condition (see Fig. 4-81).

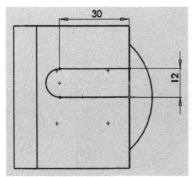

Figure 4-81 Third extruded cut

10. Create Sketch5. Select the left-hand outside face of Extrude-Thin1 as the sketch plane. Sketch two circles 20 mm apart, 20 mm above the bottom of Sketch4, which passes through the origin; the center of the left-hand circle is 30 mm from the right-hand edge (see Fig. 4-82).

11. Using these two circles, create the fourth extruded cut feature. Select Up To Next as the default End Condition.

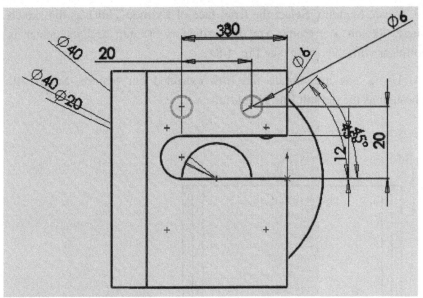

Figure 4-82 Fourth extruded cut

12. Create the first mirror feature. Mirror the two top holes about the edge passing through the origin (see Fig. 4-83).

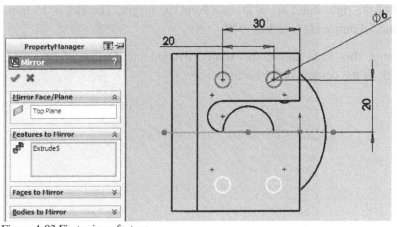

Figure 4-83 First mirror feature

13. Create Sketch6. Select the front face of Extrude-Thin1 as the sketch plane. Sketch a 3 Point Arc; the radius is 30 mm and its center is collinear with the origin (see Fig. 4-84).

14. Using this arc, create the fifth extruded cut feature. Select All Through as the default End Condition.

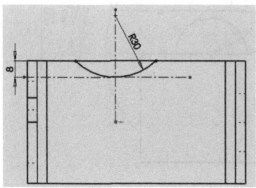

Figure 4-84 Sketch6

15. Create Sketch7. Select the front face of Extrude-Thin1 as the sketch plane. Sketch a slot. The center is 30 mm from the top of the part. The center of the right-hand arc is 20 mm from the origin. The slot is 15 mm long and 12 mm wide (see Fig. 4-85).

16. Using this slot, create the sixth extruded cut feature. Select All Through as the default End Condition.

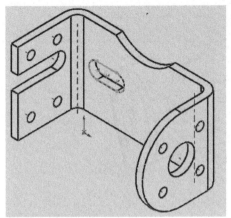

Figure 4-85 Sketch7

17. Create the second Mirror feature. Mirror the slot about the edge passing through the origin (see Fig. 4-86).

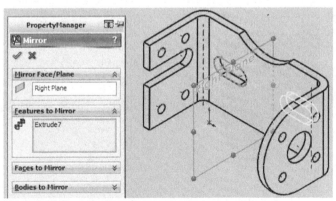

Figure 4-86 Second mirror feature

18. Use the Chamfer tool. Apply the Angle-Distance option, with an angle of 45° and a distance of 5 mm (see Fig. 4-87).

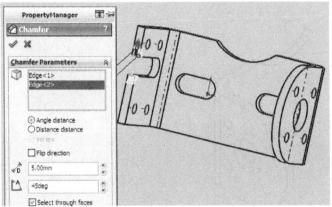

Figure 4-87 Chamfer PropertyManager

19. Assign the type of material to the part. Fig. 4-88 shows the final part and its material properties are listed.

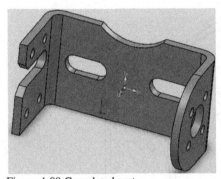

Figure 4-88 Completed part

Mass = 135.48 gm
Volume = 50176.12 mm^3
Surface area = 24647.71 mm^2
Center of mass (mm): X = 1.78, Y = -0.86, Z = -35.65

Tutorial 4.9

We will design a part in SolidWorks given the following information and using Fig. 4-89:

Units: MMGS

$A = \phi 90$, top of right-angled face and through the other two features

Material: gray cast iron

Density: 0.0072 g mm^{-3}

Part origin: middle of left-hand side

Note: All holes are full through unless otherwise noted

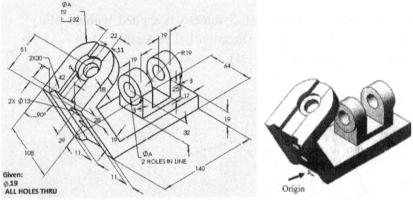

Figure 4-89 Part details

1. Create Sketch1, which is the base sketch. Select the right-hand plane. Place the origin at the bottom of the left-hand side of the sketch (see Fig. 4-90).

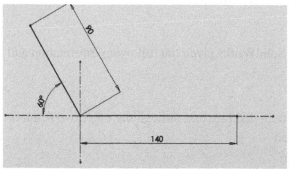

Figure 4-90 Sketch1

2. Create feature Extrude-Thin1. Symmetry is applied, with the Mid Plane as the End Condition in Direction 1. The depth is 64 mm and the thickness is 19 mm (see Fig. 4-91).

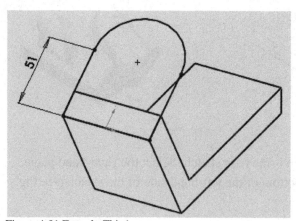

Figure 4-91 Extrude-Thin1

3. Create Sketch2. Select the top of the face of Extrude-Thin1 as the sketch plane. There are three lines and an arc (use Tangent Arc). The two opposite lines are parallel and the lines and the arc are coplanar with the top face of Extrude-Thin1 (see Fig. 4-91).
4. Use the extrusion tool on Sketch2, to create the feature Extrude1. The depth is 18 mm using Blind as the default End Condition (see Fig. 4-92).

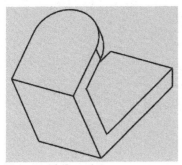

Figure 4-92 Extrude1

5. Create Sketch3. Select the right-hand plane as the sketch plane. This coincides with the part origin and cuts the part symmetrically in half, since Mid Plane was used for the extrusion of Sketch1. Sketch a rectangle of 61 mm by 31mm (51 mm – 19 mm). Note: this is actually on the right-hand plane (see Fig. 4-93).

6. Using this rectangle, create the Extrude2 feature. Select the Mid Plane as the End Condition in Direction 1, with a depth of 38 mm. Extrusion is carried out symmetrically about the Mid Plane (19 mm on both sides).

7. Create Sketch4. Select the Mid Plane as the sketch plane. Sketch a centerline for Extrude2 and mirror about it to give a rectangle of 25 mm by 29 mm. Ensure that the lines of the rectangle are either horizontal or vertical.

8. Using this rectangle, create the first extruded cut feature. Extrude on both directions by selecting Blind as the End Condition and 32 mm depth in each direction, Direction 1 and Direction 2. This creates the U-shaped feature (see Fig. 4-93).

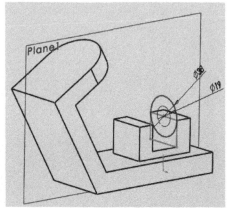

Figure 4-93 Sketch5

9. Create Sketch5. Select the inside face of Sketch3 for the sketch plane. Use the mid-point of the top line to create a circle of diameter 38 mm (see Fig. 4-93).

10. Use the Extrude Boss tool with this circle. Use Blind as the End Condition, with a depth of 19 mm.

11. Create Sketch6. Select the inside face of Sketch3 for the sketch plane. Use the mid-point of the top line to create a circle of diameter 19 mm (see Fig. 4-94).

12. Using this circle, create the second extruded cut feature. Use Blind as the End Condition, with a depth of 19 mm (see Fig. 4-94).

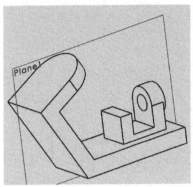

Figure 4-94 Second extruded cut

13. Create Plane3 for mirroring the extruded boss and extruded cut features. Use the Parallel Plane at Point condition to define Plane3. The Parallel Plane is the inner face of the right-hand extruded boss, while the point is the mid-point of the horizontal line of the U-shaped feature (see Fig. 4-95).

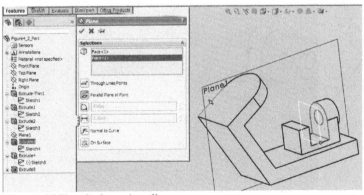

Figure 4-95 Plane3, shown in yellow

14. Mirror both the extruded boss (Extrude4) and the extruded cut (Extrude5) about Plane3. (See Fig. 4-96 for a preview and Fig. 4-97 for the mirrored features.)

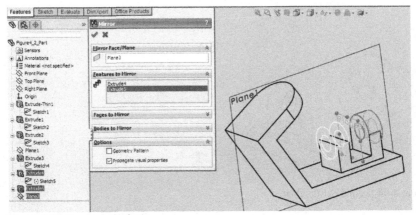

Figure 4-96 Preview of mirror

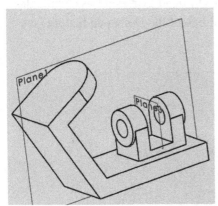

Figure 4-97 Mirrored features

15. Create Sketch7. Select the front, angled face of Extrude-Thin1 as the sketch plane. Using a centerline, and taking advantage of symmetry about it, sketch a rectangle (see Fig. 4-98).

16. Create the third extruded cut feature. Select All Through as the End Condition and the angle edge as a vector for extrusion (see Fig. 4-98).

17. Create Sketch8. Select the top face of Extrude1, which is the face with an arc, as the sketch plane. Sketch the two lines using a centerline

162

for symmetry. Apply the Convert Entities Sketch Tool to extract part of the arc needed. Take advantage of symmetry about the centerline (see Fig. 4-98).

18. Create the fourth extruded cut feature. Select Blind as the End Condition in Direction 1, with a depth of 6 mm (see Fig. 4-98).

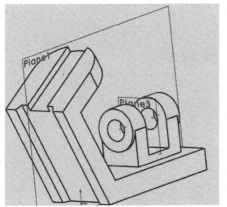

Figure 4-98 Sketch7 and Sketch8 and their extruded cuts

19. Create Sketch9. Create a center bore by selecting the top face of Extrude1 as the sketch plane (fourth extruded cut). Sketch a circle of diameter 32 mm.
20. Create the fifth extruded cut feature. Select Blind as the End Condition in Direction 1, with a depth of 9 mm (see Fig. 4-99).
21. Create Sketch10. Create a center bore by selecting the top face of Extrude1 as the sketch plane (fourth extruded cut). Sketch a circle of diameter 19 mm.
22. Create the sixth extruded cut feature. Select Blind as the End Condition in Direction 1, with a depth of 6 mm (see Fig. 4-99; see also Fig. 4-100 for the 3D view).

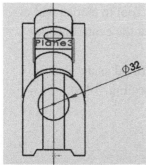

Figure 4-99 Sketch9 and Sketch10 and their extruded cuts

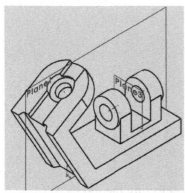

Figure 4-100 3D view

23. Create Sketch11. Select the front, angled face of Extrude-Thin1 as the sketch plane (third extruded cut). Sketch two circles 29 mm apart, each 13 mm in diameter. Apply the Convert Entities Sketch Tool to extract part of the arc needed (see Fig. 4-101).

24. Create the last extruded cut feature. Select Up To Next as the End Condition.

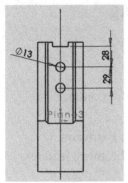

Figure 4-101 Sketch11

The completed part is shown in Fig. 4-102 and its mass properties are listed.

Figure 4-102 Final part

The final model has the following mass properties:

Density = 0.0072 g mm^{-3}

Mass = 2664.65 g

Volume = 370090.13 mm^3

Surface area = 64376.19 mm^2

Center of mass (mm): X = 0.00; Y = 39.63; Z = -41.41

Chapter 5

Advanced Part Modeling

Objectives:

When you complete this chapter you will have:

- Learnt how to create and manipulate a model's coordinate system
- Learnt how to model advanced 3D parts using Extrusion, Revolved, Lofted, Swept, Modification, Edit Features, features tools as well as using Reference Planes
- Learnt how to apply Advanced Modeling Tools
- Learnt how to create Dome, Shape features
- Learnt how to create Rib, Mounting Boss features
- Learnt how to create Multiple Bodies
- Learnt how to create Indent features

5.1 Introduction

The differences between *advanced part modeling* and *intermediate part modeling* are due to the further complexity of the sketches, with more dimensions, geometric relations, and features. This means that the modeler has to be able to deal with a significant amount of information in order to complete an advanced part model. In the first part of this chapter, we present some advanced part modeling tutorials.

Another difference between *advanced part modeling* and *part modeling* is the complexity of the features. In the second part of this chapter, we present some features that are used to alter the shape of a part. They are known as *advanced modeling tools*, and include the following:

- Dome and shape features
- Rib features
- Mounting boss features
- Multiple bodies
- Indent features

It is important to keep in mind the design intent. Take advantage of symmetric features and build in relations that would shorten design time. Maintenance of future design depends to a great extent on how well the current design has been carried out.

Advanced part modeling is one of the five categories of the Certified SolidWorks Associate (CSWA) examination on the www.solidworks. com/cswa website. One question in this chapter is taken from the CSWA category.

5.2 Advanced Part Modeling Tutorials

Tutorial 5.1: Block with hook
We will create the model shown in Fig. 5-1. We will determine the overall mass and volume of the part, and the center of mass for the model if made of 1060 alloy. Note the origin of the model.

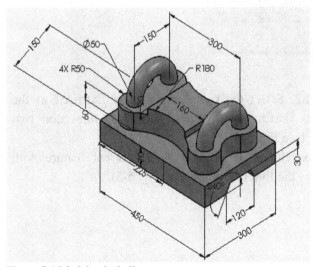

Figure 5-1 Model to be built

1. Create a new part in SolidWorks.
2. Set the document properties for the model.
3. Create Sketch1, which is 450 mm by 300 mm. Select the top plane as the sketch plane. Note that using the Center Rectangle tool is useful due to the origin of the part (see Fig. 5-2).
4. Use the Extrude Base tool. The extrusion depth is 70 mm. Extrude1 is the base feature (see Fig. 5-3).

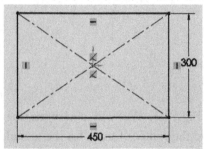

Figure 5-2 Sketch1

Figure 5-3 Extrude1

5. Create Sketch2. Select the right-hand face of Extrude1 as the sketch plane. Sketch two parallel, horizontal lines and two inclined lines (see Fig. 5-4).
6. Using this sketch, create the first extruded cut feature with Through All as the End Condition (see Fig. 5-5).

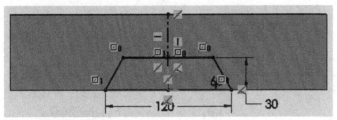

Figure 5-4 Sketch2

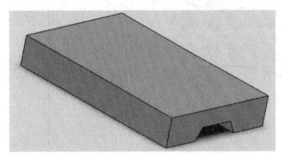

Figure 5-5 Extrude-Cut1

7. Create Sketch3. Select the top face of Extrude1 as the sketch plane. Sketch the profile shown (see Fig. 5-6).
8. Use the Extrude Boss tool on this sketch. Extrude the feature using the Blind End Condition, with a depth of 58 mm (see Fig. 5-7).

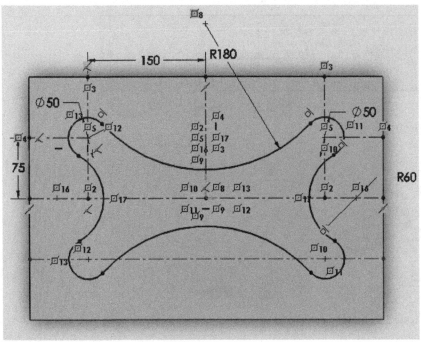

Figure 5-6 Sketch of profile on top of Extrude1

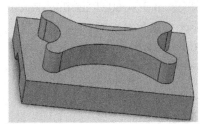

Figure 5-7 Extrusion of the sketched profile, Extrude2

9. Create Sketch4. Select the top face of Extrude2 as the sketch plane. Sketch the profile shown. This is easily done by using the Convert Entities Tool to extract the two bounding curves, drawing two vertical lines, and trimming appropriately (see Fig. 5-8).

10. Create the second extruded boss using this sketch. Extrude the feature using the Blind End Condition, with a depth of 20 mm (see Fig. 5-9).

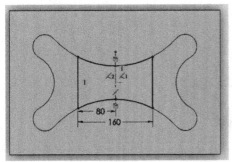

Figure 5-8 Sketch of profile on top of Extrude2

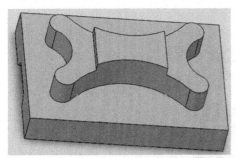

Figure 5-9 Extrusion of the sketched profile, Extrude3

11. Create Sketch5. Select the top face of Extrude2 as the sketch plane. Sketch a circle, diameter 50 mm (see Fig. 5-10).
12. Exit Sketch5.
13. Create Plane1 cutting through the center of Sketch5. Select the right-hand plane.
14. Create Sketch6. Select Plane1 as the sketch plane. Sketch an arc, radius 75 mm (see Fig. 5-11).
15. Exit Sketch6.

16. Create the first swept boss feature. Sketch5 is the profile, and Sketch6 is the path (see Fig. 5-12).

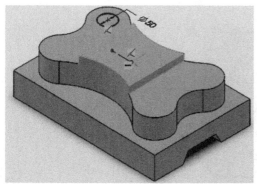

Figure 5-10 Sketch5: a circle on top of Extrude2

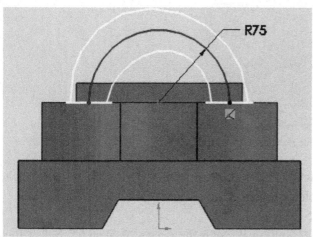

Figure 5-11 Sketch6: an arc, blue, on Plane1 on top of Extrude2

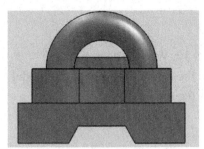

Figure 5-12 Swept feature

17. Create a mirror of the swept feature. Choose the right-hand plane to mirror (see Fig. 5-13 for the final modeled part).

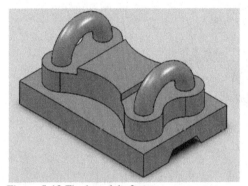

Figure 5-13 Final model of part

The mass properties are as follows:

Density = 0.003 g mm^{-3}

Mass = 37416.191 g

Volume = 13857848.552 mm^3

Surface area = 542149.374 mm^2

Center of mass (mm): X = 0.000; Y = 70.936; Z = 0.000

Tutorial 5.2: Bracket 1

We will create the model shown in Fig. 5-14. We will determine the overall mass and volume of the part, and the center of mass for the model if made of 6061 alloy. Note the origin of the model.

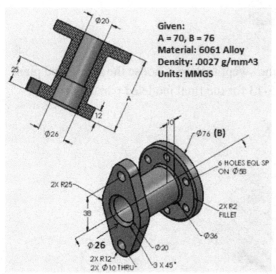

Figure 5-14 Model to be built

1. Create a new part in SolidWorks.
2. Set the document properties for the model.
3. Create Sketch1, which is made up of two circles, of diameter 76 mm and 20 mm, respectively. Select the front plane as the sketch plane (see Fig. 5-15).
4. Use the Extrude Base tool. The extrusion depth is 10 mm. Extrude1 is the base feature (see Fig. 5-16).

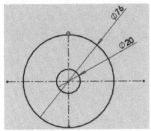

Figure 5-15 Sketch1

Figure 5-16 Extrude1

5. Create Sketch2, the profile for the first extruded cut feature. Select the front face of Extrude1 as the sketch plane. Sketch a construction circle with diameter 58 mm, and a circle with diameter of 8 mm as the seed feature for a circular pattern using the construction circle (see Fig. 5-17).

6. Using the seed circle, create an extruded cut feature. Extrude using Through All for the End Condition. Extrude2 is the first bolt hole.

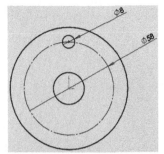

Figure 5-17 Sketch2

7. Create the circular pattern. Select a center axis for the pattern axis. Enter 6 as the number of instances, and Extrude2 as the Feature to Pattern (see Fig. 5-18).

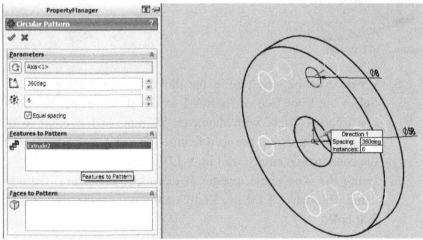

Figure 5-18 Circular pattern with the seed hole

8. Create Sketch3, which is made up of two circles, of diameter 36 mm and 20 mm, respectively. Select the front face of Extrude1 as the sketch plane.

9. Use the Extrude Boss tool with the two circles. The extrusion depth is 48 mm. Extrude3 is the boss feature, which is a hollow cylinder.

10. Create Sketch4. Select the front circular face of Extrude3 as the sketch plane. The profile of Sketch4 is shown in Fig. 5-19.

11. Using the sketch, create the second extruded boss. The extrusion depth is 12 mm. Extrude4 is the boss feature. See Fig. 5-20 for the part so far.

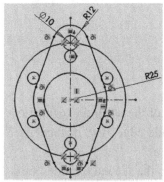

Figure 5-19 Profile of Sketch4

Figure 5-20 Part with Extrude4

12. Create Sketch5. Select the front face of Extrude4 as the sketch plane. Sketch a circle with a diameter of 26 mm (see Fig. 5-21).

177

13. Using this circle, create the second extruded cut feature. The extrusion depth is 25 mm.

Figure 5-21 Part after the second extruded cut

14. Create a chamfer. Select the top face of the second extruded cut as the sketch plane. Chamfer the inside edge. The distance is 3 mm and the angle is 45° (see Fig. 5-22a).
15. Create the fillets. Select two circles on the second extruded boss. The distance is 3 mm and the angle is 45° (see Fig. 5-22b).
16. Assign the material.
17. Calculate the overall mass of the part.
18. Locate the center of mass relative to the part's origin.

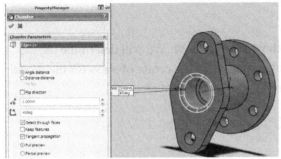

(a) Preview of chamfer PropertyManager

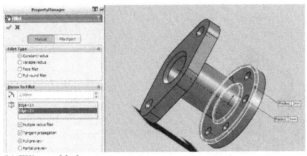

(h) Fillets added

Figure 5-22 PropertyManagers for chamfer and fillets

The final part and its material properties are shown in Fig. 5-23.

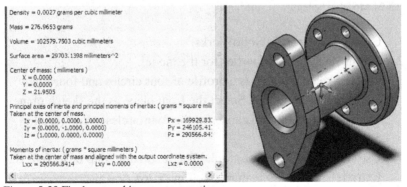

Figure 5-23 Final part and its mass properties

Tutorial 5.3: Bracket 2

We will create the model shown in Fig. 5-24. We will determine the overall mass and volume of the part, and the center of mass for the model if made of 1060 alloy. Note the origin of the model.

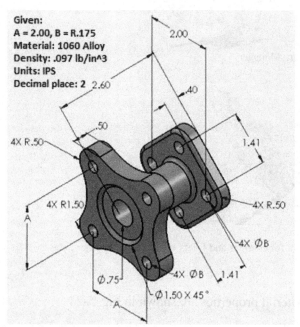

Given:
A = 2.00, B = R.175
Material: 1060 Alloy
Density: .097 lb/in^3
Units: IPS
Decimal place: 2

Figure 5-24 Model to be built

1. Create a new part in SolidWorks.
2. Set the document properties for the model.
3. Create Sketch1, which is a profile of four circles and four lines connected by four arcs. Select the front plane as the sketch plane. The center-to-center distance for the four circles is 2 in (see Fig. 5-25).
4. Use the extruded base tool. The extrusion depth is 0.4 in. Extrude1 is the base feature (see Fig. 5-26).

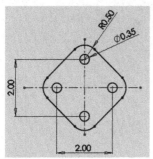

Figure 5-25 Sketch1

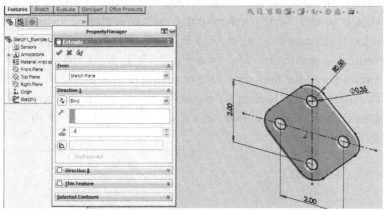

Figure 5-26 Preview of Extrude1

5. Create Sketch2, which is a circle of diameter 1.1 in. Select the front face of Extrude1 as the sketch plane (see Fig. 5-27).

6. Use the extruded boss tool on this circle. The extrusion depth is 1.7 in [= 2.60 in − (0.50 in + 0.4 in)]. The extruded boss is Extrude2.

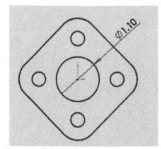

Figure 5-27 Sketch2

7. Create Sketch3. Select the front circular face of Extrude2 as the sketch plane. Sketch a horizontal and a vertical construction lines. Then create a circle, 0.35 in diameter. Mirror it about the vertical construction line (see Fig. 5-28). Next, mirror both circles about the horizontal line to give four circles; then include two arcs in the horizontal direction, two arcs in the vertical direction and four corner arcs (see Figs 5-29 and 5-30 for different views).

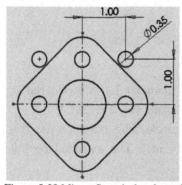

Figure 5-28 Mirror first circle, about the vertical construction line

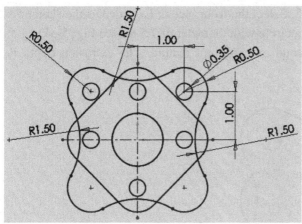

Figure 5-29 Sketch3

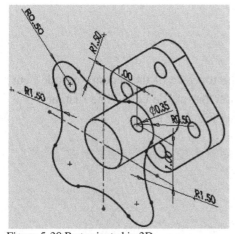

Figure 5-30 Part oriented in 3D

8. Create the second extruded boss. The extrusion depth is 0.50 in. The extruded boss is Extrude3.

9. Create Sketch4. Select the front face of Extrude3 as the sketch plane. Sketch a circle with diameter of 0.75 in (see Fig. 5-31).

10. Using this circle, create the first extruded cut feature. Select Through All as the End Condition.

11. Create Sketch5. Select the front face of Extrude3 as the sketch plane. Sketch a circle with diameter of 1.5 in (see Fig. 5-31).

12. Create the second extruded cut feature. The extrusion depth is 0.1 in.

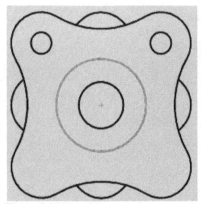

Figure 5-31 Sketch5

13. Create a chamfer. Select the top face of the second extruded cut as the sketch plane. Chamfer the inside edge. The distance is 3 mm and the angle is 45° (see Fig. 5-32).

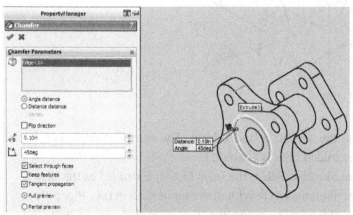

Figure 5-32 Chamfer preview

14. Create a fillet. Select the top circle of the second extruded boss. The distance is 0.1 in and the angle is 45° (see Fig. 5-33).

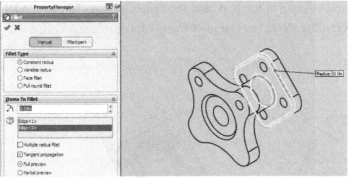

Figure 5-33 Fillet preview

15. Assign the material.
16. Calculate the overall mass of the part.
17. Locate the center of mass relative to the part's origin.

The final part and the material properties are shown in Fig. 5-34.

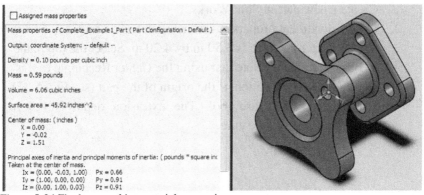

Figure 5-34 Final part and its material properties

185

Tutorial 5.4: Inclined Block

We will create the model shown in Fig. 5-35. We will determine the overall mass and volume of the part, and the center of mass for the model if made of 1060 alloy. Note the origin of the model.

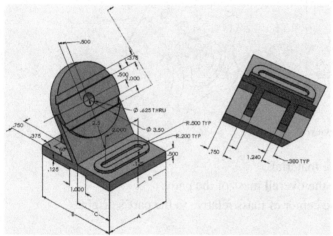

Figure 5-35 Model to be built

1. Create a new part in SolidWorks.
2. Set the document properties for the model.
3. Create Sketch1, which is 3.50 in by 4.20 in. Select the top plane as the sketch plane. Note that using the Center Rectangle tool is useful due to the position of the origin of the part (see Fig. 5-36).
4. Use the Extrude Base tool. The extrusion depth is 0.5 in. Extrude1 is the base feature.

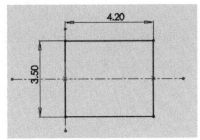

Figure 5-36 Sketch1

5. Create Sketch2. Select the top face of Extrude1 as the sketch plane. Sketch a center line (see Fig. 5-37).
6. Create Plane1, which is inclined at 60° to Extrude1 (see Fig. 5-37).

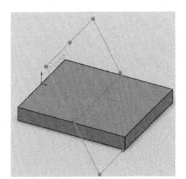

Figure 5-37 Plane1

7. Create Sketch3. Select Plane1 as the sketch plane. Sketch two equal vertical lines. Sketch an arc joining the ends of the two lines using the Tangent Arc Tool (see Fig. 5-38).
8. Create the first extruded boss. The extrusion depth is 0.26 in [= 0.50 in − 0.24 in] (see Fig. 5-39). The extruded boss is Extrude2.

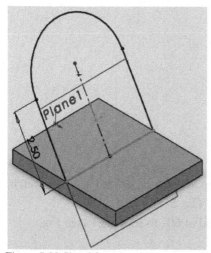

Figure 5-38 Sketch3

Figure 5-39 First extruded boss

9. Create Sketch4. Select the upper face of Extrude2 as the sketch plane. Sketch a circle coincident and co-radial to the arc of Extrude2. Sketch an arc joining the two vertical lines using the Tangent Arc Tool (see Fig. 5-40).

10. Using the circle, create the second extruded boss. The extrusion depth is 0.24 in.

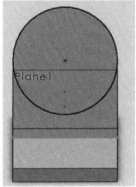

Figure 5-40 Sketch4

11. Create Sketch5, the profile for the extruded cut feature. Select the upper face of Extrude3 as the sketch plane. Sketch four horizontal lines, as chords, across the circle. Trim parts of the circle (see Fig. 5-41).
12. Use the Extruded Cut tool to create two grooves between the four chords. Extrude using the Blind for End Condition, to a depth of 0.125 in (see Fig. 5-42).

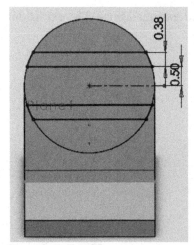

Figure 5-41 Sketch5

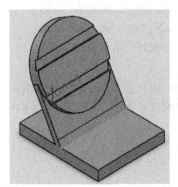

Figure 5-42 Extruded cut

13. Create Sketch6. Select the lower (or rear) face of Extrude2 as the sketch plane. Sketch a circle with a diameter equal to 0.7 in (see Fig. 5-43).

14. Create the third extruded boss. The extrusion depth is 0.20 in [= 0.70 in – 0.50 in].

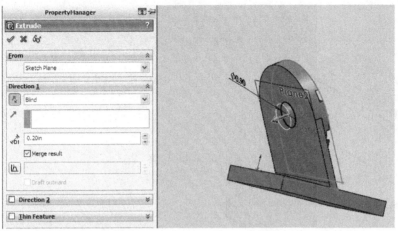

Figure 5-43 Rear boss added

15. Create Sketch7. Select the upper face of Extrude2 as the sketch plane. Sketch a circle of 0.625 in diameter (see Fig. 5-44).
16. Use the Extruded Cut tool on this circle. Extrude by selecting Through All for the End Condition. See Fig. 5-45 for a preview, and Fig. 5-46 for the partly finished part.

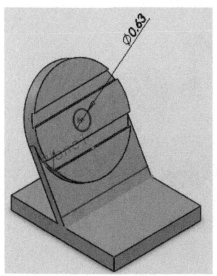

Figure 5-44 Adding a circle, to the upper face of the incline

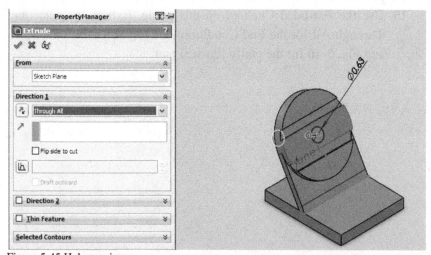

Figure 5-45 Hole preview

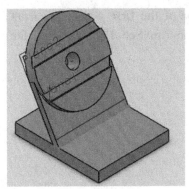

Figure 5-46 Partly finished part

17. Create a rib. Select the lower (or rear) face of Extrude2 as the sketch plane. Select Blind as the End Condition, and a depth of 0.38 in (see Fig. 5-47).

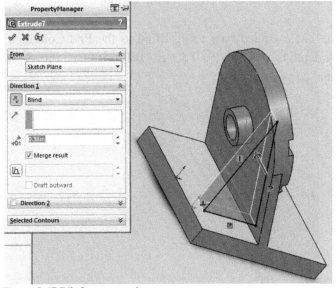

Figure 5-47 Rib feature preview

18. Create the second rib as a mirror of the first. Select the front plane as the mirror plane for the rib. See Fig. 5-48 for the mirrored rib.

Figure 5-48 Mirrored rib

19. Create Sketch8. Select the bottom of Extrude1 as the sketch plane. Sketch two slots as profiles for the extruded boss (see Fig. 5-49).
20. Create the last extruded boss using Blind as the End Condition through a depth of 0.625 in [= 0.50 in + 0.125 in] (see Fig. 5-50 for the final part).

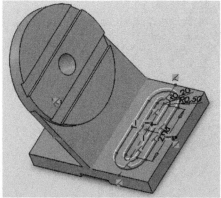

Figure 5-49 Slot profiles for extruded boss

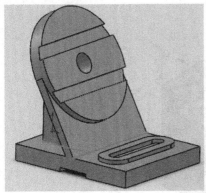

Figure 5-50 Final part.

The mass properties are as follows:

Density = 0.098 lb in^{-3}

Mass = 1.365 lb

Volume = 13.990 in^3

Surface area = 79.772 in^2

Center of mass (in): X = 1.595; Y = 0.692; Z = 0.000

Tutorial 5.5: Generalized Block

This example (Fig. 5-51) was taken from the SolidWorks website, www.solidworks.com/cswa, as an example of an advanced part for the Certified SolidWorks Associate (CSWA) examination. The model has thirteen features and twelve sketches. We will build this part. We will calculate the overall mass and locate the center of mass of the given model.

Given: A = 63, B = 50, C = 100. Material: Copper. Units: MMGS. Density = 0.0089 g mm^{-3}. All holes are through all.

It is worth noting where the origin is located in each example.

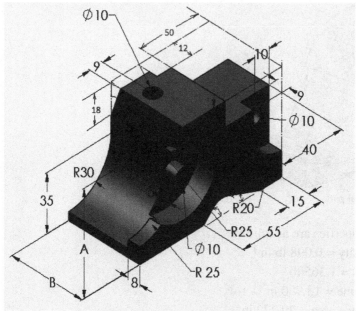

Figure 5-51 Model to be built

1. Create a new part in SolidWorks.
2. Set the document properties for the model.
3. Create Sketch1, which is 100 mm by 63 mm. Select the right-hand plane as the sketch plane. The part origin is at the left-hand corner of the sketch (see Fig. 5-52).
4. Use the Extrude Base tool. The extrusion depth is 50 mm. Extrude1 is the base feature (see Fig. 5-53).

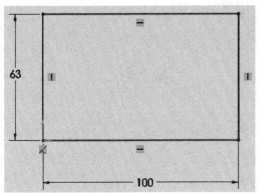

Figure 5-52 Sketch1

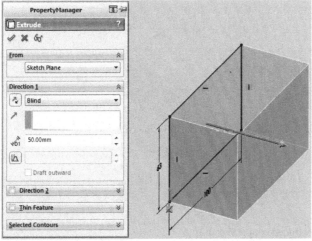

Figure 5-53 Preview for Extrude1

5. Create Sketch2. Select the rear face of Extrude1 as the sketch plane. Sketch two parallel, vertical lines, of different lengths, and use the Tangent Arc tool to connect their lower ends. Sketch a horizontal line between their top ends (see Fig. 5-54).

6. Use the Extruded Cut tool with this sketch. Offset the extruded cut by 8 mm. Choose the outer face as Face<2> and an edge, Edge<1> for the direction (see Fig. 5-55).

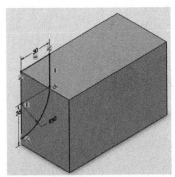

Figure 5-54 Sketch2

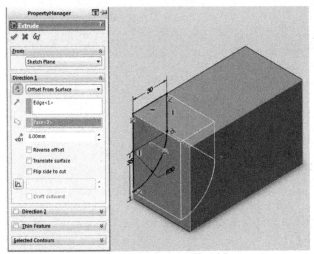

Figure 5-55 Preview of extruded cut on Sketch2

7. Create Sketch3. Select the rear plane of Extrude1 as the sketch plane (see Fig. 5-56).
8. Create a second extruded cut. Extrude using the Through All condition (see Fig. 5-57).

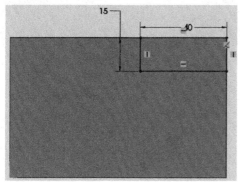

Figure 5-56 Sketch3

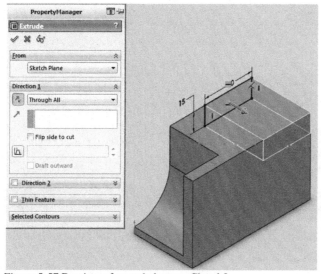

Figure 5-57 Preview of extruded cut on Sketch3

9. Create Sketch4. Select the top face of Extrude1 as the sketch plane (see Fig. 5-58).

10. Create the third extruded cut. Extrude cut using the Through All condition (see Fig. 5-59).

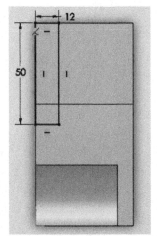

Figure 5-58 Sketch4

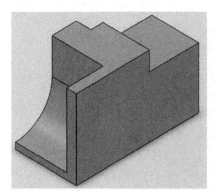

Figure 5-59 After extruded cut on Sketch4

11. Create Sketch5. Select the front face of Extrude1 as the sketch plane. Sketch two parallel, vertical lines of different lengths, and use the Tangent Arc tool to connect their lower ends. Sketch a horizontal line between their top ends (see Fig. 5-60).

12. Use the Extruded Cut tool with this sketch. Extrude cut by 9 mm (see Fig. 5-61). This extruded cut is Extrude5.

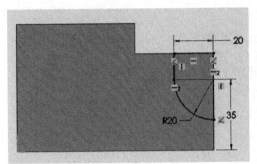

Figure 5-60 Sketch5

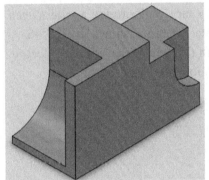

Figure 5-61 After extruded cut on Sketch5

13. Create Sketch6. Select the front face of Extrude5 as the sketch plane. Sketch a circle of diameter 10 mm and place it 10 mm from the right-hand edge and top (see Fig. 5-62).

14. Use the Extruded Cut tool on this circle. Extrude cut selecting Through All for the End Condition (see Fig. 5-62).

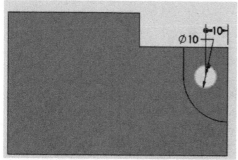

Figure 5-62 Extruded cut on Sketch6

15. Create Sketch7. Select the top face of Extrude1 as the sketch plane. Sketch a circle of diameter 10 mm as shown in Fig. 5-63.
16. Use the Extruded Cut tool with this circle. Extrude cut selecting Through All for the End Condition (see Fig. 5-63).

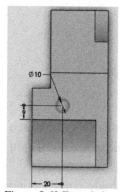

Figure 5-63 Extruded cut on Sketch7

17. Create Sketch8. Select the front face of Extrude1 as the sketch plane. Sketch a triangle and insert a tangent arc (fillet the two intersecting lines at the top vertex). See Fig. 5-64.
18. Use the Extruded Cut tool on this sketch. Extrude cut selecting Through All for the End Condition.

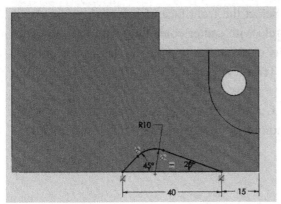

Figure 5-64 Sketch8

19. Create Sketch9. Select the front face of Extrude1 as the sketch plane. Select Hidden Lines Visible. Sketch two construction circles, concentric, with their centers placed at the end of the arc in Sketch2. Sketch an arc using the 3 Point Arc sketch tool. Its center is at the same height as the centers of the construction circles. Complete the sketch by inserting two vertical lines and one horizontal line at the top to create a closed shape (see Fig. 5-65).

20. Use the Extruded Cut tool on this shape. Extrude cut selecting Through All for the End Condition.

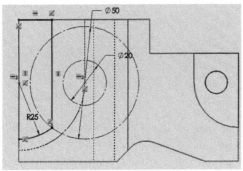

Figure 5-65 Sketch9

21. Create Sketch10. Select the front face of Extrude1 as the sketch plane. Sketch a circle with center coinciding with the end of the arc in Sketch2; this circle is the same as the larger construction circle of Sketch9. Trim it appropriately and insert a vertical line to close the shape (see Fig. 5-66).

22. Use the Extruded Cut tool on this shape. Extrude cut selecting the Blind End Condition, with a depth of 13 mm (see Fig. 5-67).

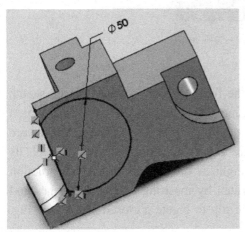

Figure 5-66 Sketch10

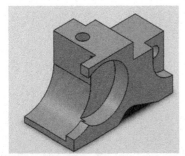

Figure 5-67 After extruded cut on Sketch10

23. Create Sketch11. Select the top face of Extrude1 as the sketch plane. Sketch the rectangle to trim the part. See Fig. 5-68.

24. Use the Extruded Cut tool with this rectangle. Extrude cut selecting the Blind End Condition, to a depth of 13 mm (see Fig. 5-69).

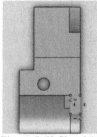

Figure 5-68 Sketch11

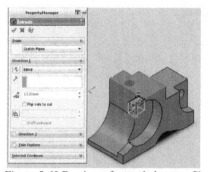

Figure 5-69 Preview of extruded cut on Sketch11

25. Create Sketch12. Select the front face of Extrude1 as the sketch plane. Sketch a circle. Trim it and extract the left-hand boundary features (line/arc). See Fig. 5-70.

26. Use the Extrude Boss tool on this trimmed circle, selecting the Blind End Condition, to a depth of 5 mm. See Fig. 5-71 and Fig. 5-72 for a 3D view.

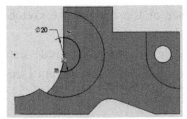

Figure 5-70 Sketch12

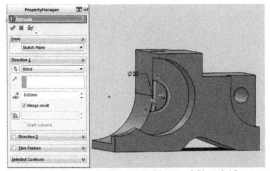

Figure 5-71 Preview of extruded boss of Sketch12

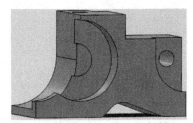

Figure 5-72 3D view of extruded boss of Sketch12

27. Create Sketch13. Select the right-hand face of Extrude1 as the sketch plane. Sketch a circle (see Fig. 5-73).

28. Use the Extrude Boss tool on this circle, selecting the Blind End Condition, to a depth of up to the outer face (Fig. 5-74).

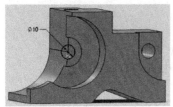

Figure 5-73 Sketch13

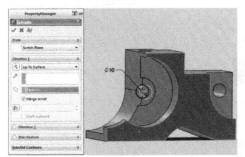

Figure 5-74 Preview of extruded boss on Sketch13

29. Create a chamfer. Select the rear face of Extrude1 as the sketch plane. The distance is 18 mm and the angle is 20° (see Fig. 5-75).

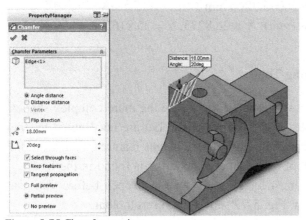

Figure 5-75 Chamfer preview

207

The final part is shown in Fig. 5-76.

30. Assign the material.
31. Calculate the overall mass of the part.
32. Locate the center of mass relative to the part's origin.

Figure 5-76 Final modeled part

The mass properties are as follows:

Density = 0.0089 g mm^{-3}

Mass = 1279.4834 g

Volume = 143762.1768 mm^3

Surface area = 26095.3206 mm^2

Center of mass (mm): X = 26.8275, Y = 25.8163, Z = -56.3862

5.3 Advanced Modeling Tools

Advanced modeling tools are needed to create the more complex shapes encountered in practice. These tools include, but are not limited to, the following: Dome, Shape, Mounting boss, Indent features, and Multiple Bodies. The features tool for advanced modeling is found in the Insert > Features toolbar. You must have a part document open before you can access the Insert toolbar. Fig. 5-77 shows the Insert toolbar.

The Insert toolbar is very important because not only does it have the features tools, it also has the Surface, Sheet Metal, Weldments, and Molds toolbars, which are essential for the various applications found in industry. The Insert toolbar is the *gateway* to a number of relevant and practical functions.

Notice the other useful tools on this toolbar, such as Split and Move/Copy, which are useful in creating solid bodies, as well as Boss/Base, Cut, and 3D Sketch that are also useful for modeling.

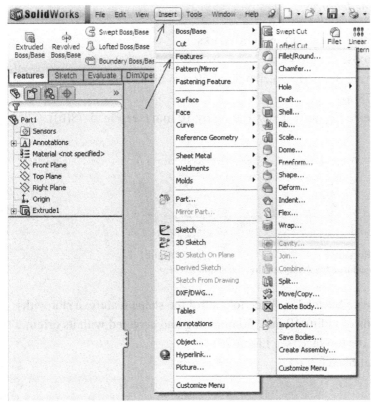

Figure 5-77 The Insert toolbar

Tool which are used to alter the shape of a part are known as advanced modeling tools. Dome features are covered in Section 3.2.1. The rest are the subject of the remaining part of this chapter and include the following:

- Shape features
- Rib features
- Mounting boss features
- Indent features

5.3.1. *Shape features*

1. Create a new part in SolidWorks.
2. Choose the top view.
3. Create Sketch1, a rectangle of 60 mm by 45 mm (see Fig. 5-78(a)).
4. Create Feature1, by extruding by 10 mm (see Fig. 5-78(b)).

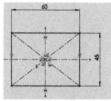 

(a) Sketch1, Rectangle   (b) Feature1

Figure 5-78 Feature1 for demonstrating the shape feature

5. Create Sketch2, a shape to control the shape feature: a slot with ends of radius 10 mm, 30 mm long and centered with its origin on the top face (see Fig. 5-79).
6. Exit Sketch2.

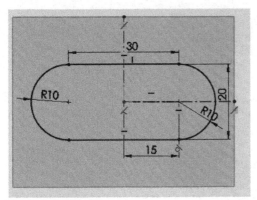

Figure 5-79 Sketch2

7. Click Feature > Insert > Feature > Shape.
8. Select Shape Feature.
9. Select the top face to manipulate the Shape tool (see Fig. 5-80(a)).
10. Select Sketch2 to constrain the shape feature (see Fig. 5-80(a)).
11. Select Preview.
12. Select the Controls tab (see Fig. 5-80(b)).
13. In the Gains area, move the *slider* to a positive value for Pressure to *inflate* the feature around Sketch2, or a negative value to *deflate* it. Similarly, change the Curve Influence.
14. In the Characteristics area, move the *slider* to a positive value for Stretch to *inflate* the feature around Sketch2, or a negative value to *deflate* it. Similarly, change Bend.

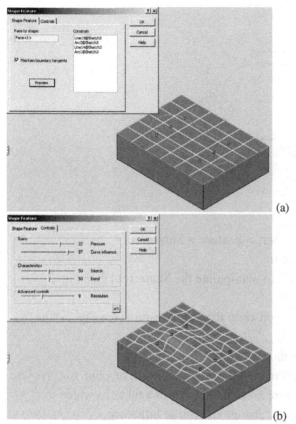

(a)

(b)

Figure 5-80 Previews of the Shape tool

15. Click OK. Fig. 5-81 shows the final shape feature.
16. Click Section View on the front view to clearly see the shape feature (see Fig. 5-82).

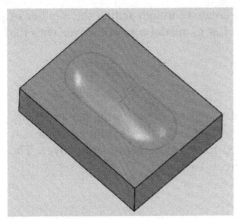

Figure 5-81 Shape feature created

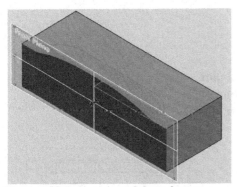

Figure 5-82 Section view of shape feature

5.3.2. *Mounting boss features*

A mounting boss feature is a fastening feature, such as a pin or a hole for a pin. Fastening features are used to mount other features onto plastic products. They streamline the creation of common features for plastic and sheet metal parts. There are a number of SolidWorks fastening features: Mounting Boss, Snap Hook, Snap Hook Groove, Vent, and Lip/Groove.

Different mounting bosses can be created through setting the number of fins and choosing a hole or a pin. Let us model a plastic cover with the mounting boss feature.

Problem description

Figure 5-83 shows the model of a plastic cover, with the necessary dimensions. The two pins inside will be created as mounting bosses. A draft angle of 1° has to be applied to the side faces of the cover. The parameters of the mounting boss are as follows:

Boss diameter = 4.8 mm
Boss height = 14 mm
Draft angle of main boss = 2°
Height of fins = 12 mm
Width of fins = 1.5 mm
Length of fins = 4 mm
Draft angle of fins = 1°
Number of fins = 4
Diameter of inside hole = 2 mm
Height of inside hole = 5 mm
Draft angle of inside hole = 1°

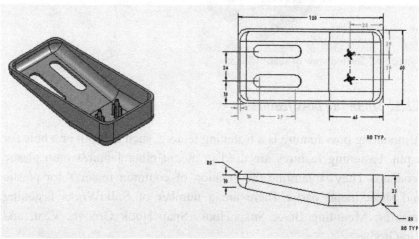

Figure 5-83 Model of a plastic cover

Sketch

1. Create a new part in SolidWorks.
2. Choose the top view.
3. Create Sketch1, a profile made up of two horizontal lines, two vertical lines, and an inclined line (see Fig. 5-84).

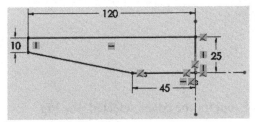

Figure 5-84 Sketch1

Extrude

4. Create Feature1 by extruding by 60 mm (see Fig. 5-85).

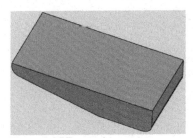

Figure 5-85 Feature1

Shell

5. Click Shell.
6. Select the top of Feature1.
7. Enter a thickness of 1 mm.
8. Click OK (see Fig. 5-86 for the shell feature).

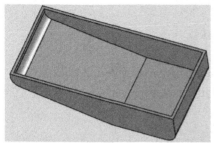

Figure 5-86 Shell feature

Fillet

9. Create fillets, of radius 8 mm for the edges selected (see Fig. 5-87).

This plastic cover will be also be used as the starting point for the models in the following sections, on ribs and indents.

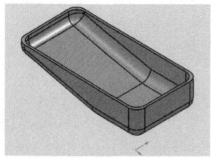

Figure 5-87 Fillets added to part

First slot

10. Create a slot, 30 mm between the centers of the end of the arcs, with radius 5 mm (see Fig. 5-88).

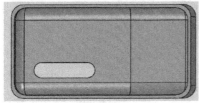

Figure 5-88 First slot added

Pattern of slots

11. Click the Linear Pattern tool (see Fig. 5-89).

12. Select a vertical line as Direction 1.

13. Set the distance between the centerlines of the two slots to 24 mm.

14. Select the number of instances to 2.

15. Select Extrude2 as the Feature to Pattern.

16. Click OK to complete the pattern of slots.

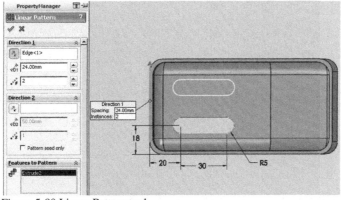

Figure 5-89 Linear Pattern tool

Mounting boss feature

17. Invoke the Mounting Boss feature tool from Insert > Fastening Feature and select the Mounting Boss option as shown in Fig. 5-90.

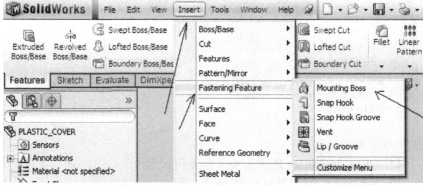

Figure 5-90 The Insert > Fastening Feature > Mounting Boss option

The Mounting Boss PropertyManager appears, as shown in Fig. 5-91(a).

18. Select the Boss rollout for the mounting boss and enter the values shown in the figure.

19. Select the Fins rollout for the mounting boss and enter the values shown in the figure.

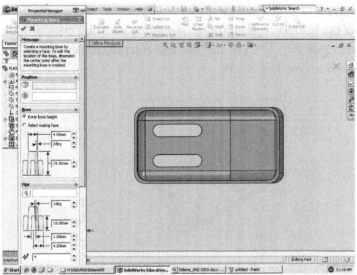

(a) Mounting Boss PropertyManager

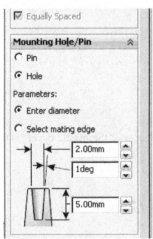

(b) Mounting Hole/Pin PropertyManager
Figure 5-91 Mounting Boss PropertyManager for setting relevant design parameters

Edit location by editing the center point

20. Select the Position rollout for the mounting boss and click a position (see Fig. 5-92) on the part and enter the dimensions of the position (see Fig. 5-93).

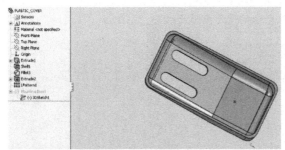

Figure 15-92 Position for mounting boss

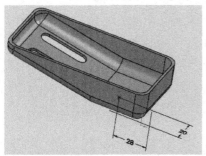

Figure 5-93 Dimensions from the center of the pin to the sides of the plastic cover (3D Sketches are involved)

Pattern of mounting bosses

21. Click Linear Pattern to activate the Linear Pattern PropertyManager (see Fig. 5-94).
22. Click an edge in the direction of the linear pattern.
23. Set the spacing to 20 mm and the number of instances to 2.
24. Click OK to complete the insertion of the mounting bosses (see Fig. 5-95 for the final part).

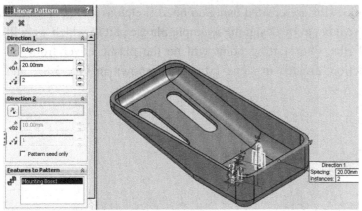

Figure 5-94 Linear Pattern PropertyManager

Figure 5-95 Final modeled part with mounting bosses

5.3.3. Rib features

Ribs are thin-walled structures that are used to increase the strength of a component, so that it can withstand an increased load.

Ribs are commonly found in home-use plastic products, such as dish racks that used for holding dishes after washing them, or in racks for storing groceries in grocery shops, supermarkets, drug marts, etc.

In SolidWorks, ribs are created using an open sketch as well as a closed sketch. We will begin by designing a simple plastic part to which we will add several ribs. Our starting point will be the plastic cover from the previous section, created in steps 1 to 9 and shown in Fig. 5-87, and again in Fig. 5-96.

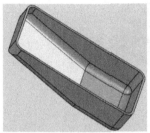

Figure 5-96 Shell feature: plastic cover

To create a rib in SolidWorks, draw a sketch and exit the sketching environment. Select the sketch from the drawing area. When the Rib tool is selected, the Rib PropertyManager pops up. This is used to enter the thickness of the rib and the extrusion direction, and to specify the confirmation corner.

Thickness

The thickness section of the Parameter rollout is used to specify the side of the sketch where the rib thickness is to be added and the thickness of the rib. The three icons for thickness are First Side, Both Sides, and Second Side, respectively, given in that order from left to right in the Thickness rollout.

Extrusion Direction

There are two options – Parallel to Sketch or Normal to Sketch. The Parallel to Sketch button is used to extrude the sketch in a direction that is *parallel* to both the sketch and the sketching plane. The Normal to

Sketch button is used to extrude the sketch in a direction that is *normal* to both the sketch and the sketching plane.

Type

The type is provided with two radio buttons – Linear and Natural. The linear option extends the sketch normal to the sketched entity direction. The natural option extends the sketch along the same direction of curvature as the sketched entities.

1. Open the existing plastic cover.
2. Sketch the required profiles: three horizontal lines and two vertical lines (see Fig. 5-97).
3. Exit the sketching environment.

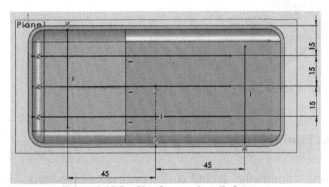

Figure 5-97 Profiles for creating rib features

4. Select the profiles from the drawing area.
5. Click the Rib tool, the Rib PropertyManager pops up (see Fig. 5-98).
6. Set the thickness as 1.50 mm.
7. Select normal as the extrusion direction.
8. Select the type as linear.
9. Confirm by clicking OK (see Fig. 5-99 for the ribbed features in the final part).

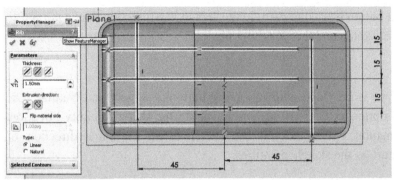

Figure 5-98 Rib PropertyManager

Figure 5-99 Final ribbed part

5.4 Indent Features

Indents are commonly included in packaging products. Industrial designers and product designers have many uses for the Indent Feature tool.

Our starting point will be the plastic cover from Section 5.3.2, created in steps 1 to 9 and shown in Fig. 5-87. The Indent Feature tool requires a target body and a tool body.

1. Open the existing plastic cover. This is the target body.
2. Create a reference plane 45 mm from the top plane.

Create another part, for the tool body. First create Sketch1 on the reference plane, a rectangle 30 mm by 20 mm, centered on the bottom of the plastic cover (see Fig. 5-100(a)). Create Extrude1 by extruding by 30 mm (see Fig. 5-100(b)). Do not check the Merge result checkbox when extruding the second part. The edges have 2.5 mm fillets.

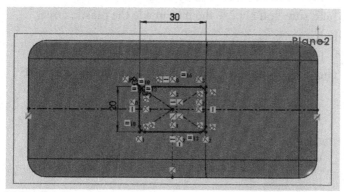

(a) Sketch1

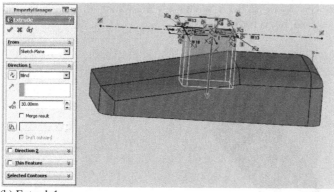

(b) Extrude1

Figure 5-100 Preview of part created by extruding Sketch1

The plastic cover is the target body while the 30 mm by 20 mm by 30 mm feature is the tool body (see Fig. 5-101). We are now ready to create the indent.

3. Click Feature > Insert > Indent. The Indent PropertyManager appears, see Fig. 5-102.
4. Select the plastic cover as the Target Body.
5. Select the new 30 mm by 20 mm by 30 mm feature as the Tool Body. Make sure you choose the part inside the plastic cover.
6. Click OK.
7. Select the top of the 30 mm by 20 mm by 30 mm feature and hide it. The result is an indented feature, as shown in Fig. 5-103.

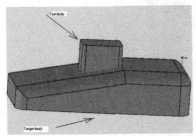

Figure 5-101 Tool body and Target body defined

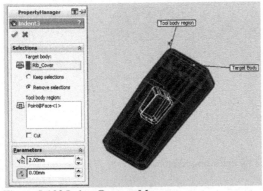

Figure 5-102 Indent PropertyManager

Figure 5-103 Final part with indent

Chapter 6

Revolved, Swept, and Lofted Parts

Objectives:

When you complete this chapter you will have:

- Learnt how to create 3D objects using Revolved Features tools
- Learnt how to use Copy Features tools
- Learnt how to create 3D objects using Swept Features tools
- Learnt how to use Draft Features tools
- Learnt how to create 3D objects using Lofted Features tools
- Learnt how to use Circular Pattern Features tools
- Learnt how to use Reference Planes
- Learnt how to use Copy Features tools

6.1 Revolved Boss/Base

The Revolved Boss/Base tool rotates a contour about an axis. It is a useful tool when modeling parts that have circular symmetry. Let us illustrate the Revolved Boss/Base tool as follows:

1. Select the front plane and create Sketch1, as shown in Fig. 6-1.
2. Click the Features tool.
3. Click the Revolved Boss/Base tool. The Revolve Property-Manager appears (see Fig. 6-2).
4. Define the revolved axis (Line1) as the long, vertical line. A real-time preview will appear.
5. Click OK to complete the revolved part, Extrude1 (see Fig. 6-3).

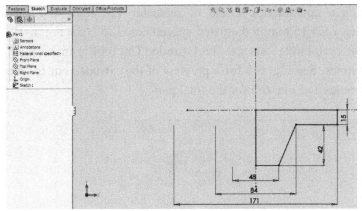

Figure 6-1 Sketch1

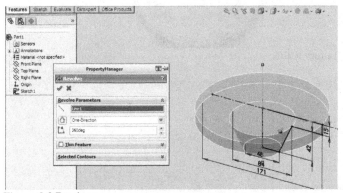

Figure 6-2 Preview

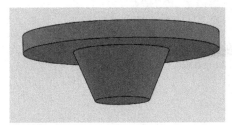

Figure 6-3 Revolved part, Extrude1

6. Click the top face of Extrude1.

7. Sketch a construction circle of diameter 126 mm, and four circles, each 15 mm in diameter, spaced out at 45° (see Fig. 6-4).

8. Select each circle, and use the Extruded Cut tool to make them into holes. See Fig. 6-5 for a preview of the extruded cut for the four holes and Fig. 6-6 for the final part.

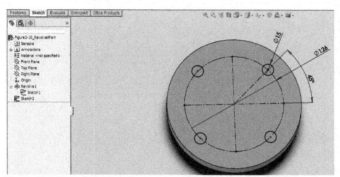

Figure 6-4 Sketch for four holes

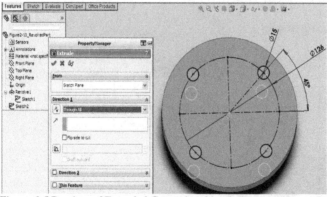

Figure 6-5 Preview of Extruded Cut tool making holes

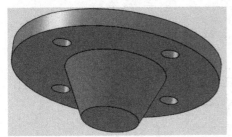

Figure 6-6 Final revolved part

6.1.1. *Practical examples*

This section gives two practical examples of revolved parts. The first is an engine cylinder, commonly used in automobile design, and the second is a pulley, commonly used in power transmission.

Engine cylinder

1. Select the front plane and create Sketch1, as shown in Fig. 6-7. Define the relations for the slots to ensure that the sketch is fully defined.
2. Click the Features tool.
3. Click the Revolved Boss/Base tool. The Revolve Property-Manager appears.
4. Define the revolved axis (Line1) as the vertical center line. A real-time preview will appear.
5. Click OK to complete the revolved part as shown in Fig. 6-8.

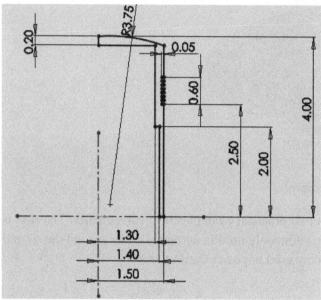

Figure 6-7 Sketch1

Figure 6-8 Final model of an engine cylinder

Pulley

1. Select the right-hand plane and create Sketch1, as shown in Fig. 6-9. This sketch is part of the cross-section through the pulley. The steps required to create the teeth in the sketch are:

    (i)   Sketch a line to the left of the vertical construction line.

    (ii)   Mirror this line about the vertical construction line.

    (iii)   Dimension the lines to be 10 mm apart.

    (iv)   Sketch the first tooth completely, and define relations so that the three horizontal lines, which are part of the tooth, are 10 mm long and the angles are 60° (see Fig. 6-10).

    (v)   Copy part of the tooth to make a second one, and ensure that the end-condition relations are met (see Fig. 6-11).

    (vi)   Mirror the two teeth about the vertical construction line (see Fig. 6-12).

    (vii)   Check that all points at the bottom are collinear, and check that the points at the top are also collinear. Also check that the sketch forms a closed loop.

2. Click the Features tool.

3. Click the Revolved Boss/Base tool. The Revolve Property-Manager appears.

4. Define the revolved axis (Line1) as the horizontal center line. A real-time preview will appear (see Fig. 6-13).

5. Click OK to complete the revolved part, as shown in Fig. 6-14.

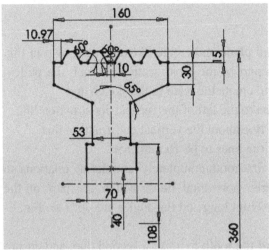

Figure 6-9 Sketch1

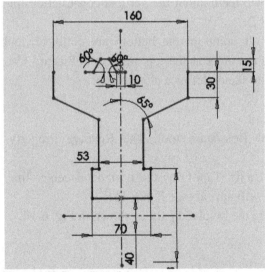

Figure 6-10 Steps (i) to (iv)

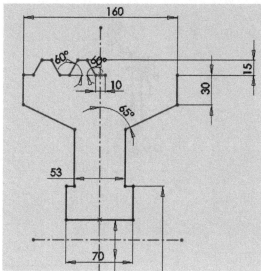

Figure 6-11 Step (v)

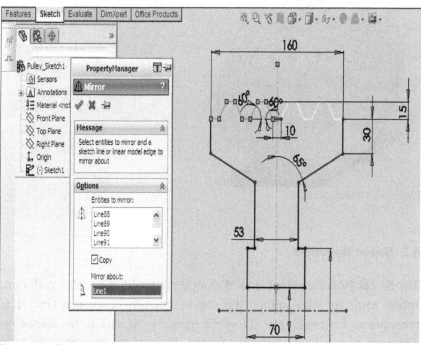

Figure 6-12 Step (vi)

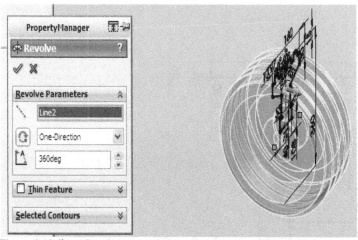

Figure 6-13 Steps 2 to 4

Figure 6-14 Pulley model

6.2 Swept Boss/Base

The Swept Boss/Base tool is used to sweep a profile through a path (arc, spline, etc). As with lofting, the shapes have to be created first. The prerequisite for creating a 3D swept model is to sketch the shapes on different perpendicular planes. In the illustration presented here, a hexagon is the *profile* while a spline is the *path*, as shown in Fig. 6-15.

1. Select the front plane from the CommandManager.
2. Click Sketch from the CommandManager or from the Context toolbar.
3. Create Sketch1 (a spline with one end at with the origin of the hexagon that will be created later).
4. Exit sketch mode.
5. From the CommandManager, click Features > Reference Geometry > Plane.
6. Select the Normal to Curve option.
7. Click the endpoint of the spline, Sketch1.
8. Click the spline, Sketch1, from the graphics window.
9. Click OK to create Plane1. This plane is normal to Sketch1.
10. Right-click Plane1 and create Sketch2 (a hexagon with sides of 0.5 in).
11. Exit sketch mode.
12. Click the Swept Boss/Base tool. The Sweep PropertyManager appears (see Fig. 6-16).
13. Right-click the Profile and Path box.
14. Click the hexagon as the profile and the spline as the path. A real-time preview will appear.
15. Click OK to complete the swept cut part, which is now hollow (see Fig. 6-17). Hide the planes.

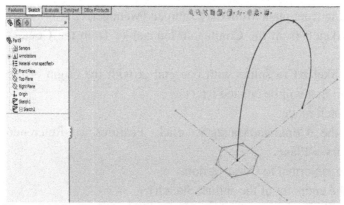

Figure 6-15 Profile (hexagon) and path (spline)

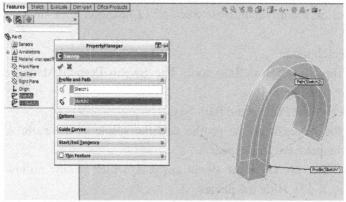

Figure 6-16 Preview and Sweep PropertyManager

Figure 6-17 3D swept model

6.2.1. Practical examples

Commonly used swept parts in industrial applications are O-rings, springs, and threaded parts (e.g. nuts and bolts, which can be found in the SolidWorks library).

Spring

Springs are modeled using a helix as the path, and the profile is a circle, which is swept through the path.

1. Select the top plane. Create Sketch1 (a circle of diameter 0.5 in) with center at the origin (see Fig. 6-18).
2. Select the isometric orientation.
3. Click on Features > Curves > Helix and Spiral (or Insert > Curves > Helix and Spiral). The Helix/Spiral dialog appears.
4. Define the helix by Height (1.75 in) and Revolution (6), and check Clockwise (see Fig. 6-19).
5. From the CommandManager, click Features > Reference Geometry > Plane.
6. Select the Normal to Curve option.
7. Click the endpoint of the helix, on Sketch1.
8. Click the helix and Sketch1 from the graphics window.
9. Click OK to create Plane1 (see Fig. 6-20). Plane1 is normal to Sketch1.
10. Right-click Plane1 and create Sketch2 (a circle of diameter 0.125 in), as shown in Fig. 6-21.
11. Exit sketch mode.
12. Click the Swept Boss/Base tool. The Sweep PropertyManager appears.
13. Right-click the Profile and Path box.
14. Click the circle as the profile and the helix as the path. It is best to do this using the FeatureManager. A real-time preview will appear (see Fig. 6-22(a)).

15. Click **OK** to complete the swept cut part, which is hollow, as shown in Fig. 6-22(b). Hide the planes.

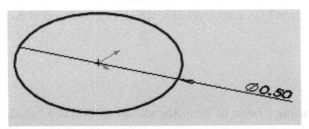

Figure 6-18 Sketch1

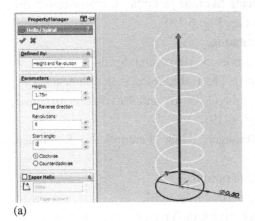

(a)

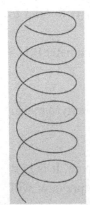

(b) Figure 6-19 Helix/Spiral PropertyManager

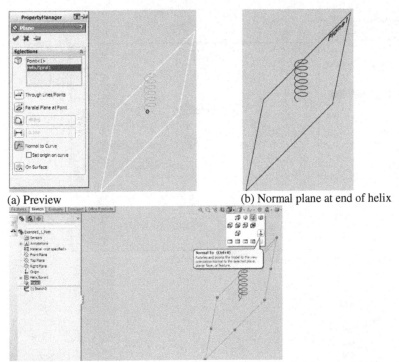

(a) Preview   (b) Normal plane at end of helix

(c) Rotating/zooming the model using the view orientation that is normal to the selected plane tool

Figure 6-20 Defining the normal plane at the endpoint of the helix

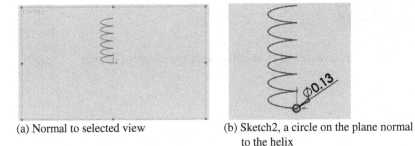

(a) Normal to selected view   (b) Sketch2, a circle on the plane normal
                                              to the helix

Figure 6-21 Circle sketched on a plane normal to the end of the helix

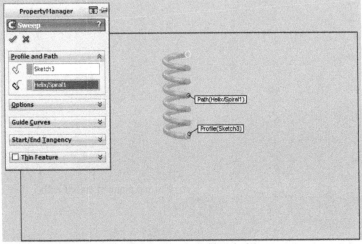

(a) Preview using path (helix/spiral) and profile (circle)

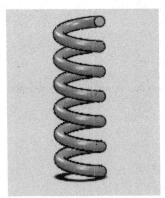

(b) Spring element

Figure 6-22 Swept Boss/Base tool for modeling a mechanical spring element

O-Ring

The procedure for modeling an O-ring is similar to that of a spring except that the path is now a circle (say, on the top plane) while the profile remains a circle, which is on a perpendicular plane (the front plane). As an illustration, let us model an O-ring with a circular path of diameter 5 in and cross-section of diameter 0.125 in.

1. Select the top plane. Create Sketch1 (a circle of diameter 5 in) with center at the origin, as shown in Fig. 6-23.
2. Select the Isometric orientation.
3. Exit Sketch mode.
4. From the CommandManager click Features > Reference Geometry > Plane.
5. Select the Normal to Curve option.
6. Click Sketch1.
7. Click OK to create Plane1. Plane1 is normal to Sketch1.
8. Right-click Plane1 and create Sketch2 (a circle of diameter 0.125 in), as shown in Fig. 6-24.
9. Add a piercing relation by clicking on the center point of the small circle, then hold the Ctrl key down. Click somewhere on the circumference of the large circle and release the Ctrl key. Right-click Make Pierce from the dialog. Click OK (see Fig. 6-25).
10. Exit sketch mode.
11. Click the Swept Boss/Base tool. The Sweep PropertyManager appears.
12. Right-click the Profile and Path box.
13. Click the small circle (Sketch2) as the profile and the large circle (Sketch1) as the path, using FeatureManager. A real-time preview will appear, as shown in Fig. 6-26.
14. Click OK to complete the swept cut part, which is hollow, as shown in Fig. 6-27. Hide the planes.

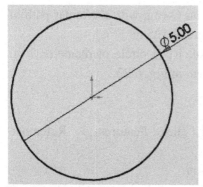

Figure 6-23 Path definition

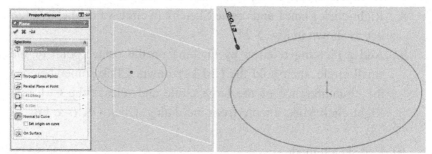

Figure 6-24 Plane definition with Profile (small circle) and path (large circle)

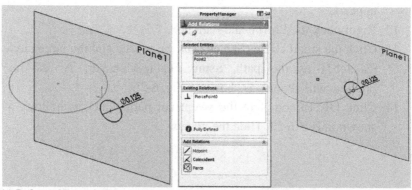

(a) Before adding relation   (b) After adding relation

Figure 6-25 Adding a piercing relation

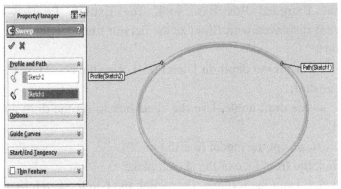

Figure 6-26 Sweep PropertyManager and O-ring preview

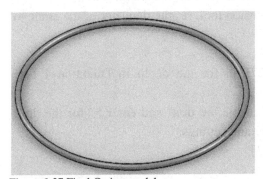

Figure 6-27 Final O-ring model

Threaded cap

Threaded parts are modeled using a helix as the path and a shape of interest swept through the path, as the profile. As a small project, let us first model a cap with a draft angle. Then, we will add a trapezoidal profile as an internal thread with a helical path.

Cap without internal thread
1. Select the front plane. Create Sketch1 (a circle of diameter 5 in) with center at the origin.

2. Click the Extruded Boss/Base tool. The Extrude Property-Manager is displayed with Blind as the default End Condition in Direction 1.

3. Enter 1.725 in for the depth in Direction 1. Use the checkbox to enable the draft.

4. Enter 5° as the draft angle. Use the checkbox to set the draft as outward.

5. Click OK to accept the model (see Fig. 6-28).

6. Right-click the front face of the sketch plane.

7. Click Sketch from the toolbar. Then click the Circle Sketch tool to create a circle centered at the origin (see Fig. 6-29).

8. Using the Smart Dimension tool, set the diameter of the circle to 3.875 in.

9. Click Extrude cut.

10. Enter a depth of 0.275 in for the depth in Direction 1 (see preview in Fig. 6-30).

11. Use the checkbox to enable the draft and enter 5° for the draft angle, accepting the default settings.

12. Click OK (see Fig. 6-31).

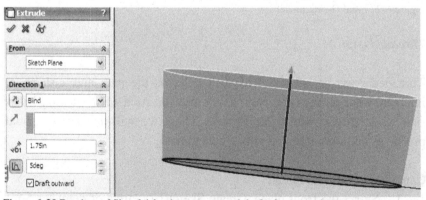

Figure 6-28 Preview of Sketch1 having an outward draft after extrusion

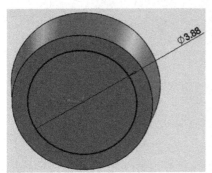

Figure 6-29 Sketch2, a circle for an extruded cut

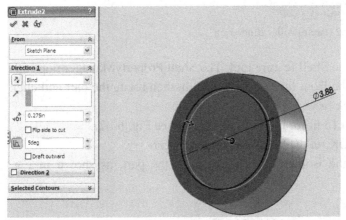

Figure 6-30 Preview for extruded cut using Sketch2 (circle)

Figure 6-31 Sketch2 (circle) with extruded cut

13. Click the Shell feature tool. The Shell PropertyManager pops up.
14. Click the front face of the front cut, then rotate the part and click the back face.
15. Enter 0.15 in for the shell thickness (see Fig. 6-32).
16. Click OK on the Shell PropertyManager.
17. Click the isometric view and save the part, as shown in Fig. 6-33.

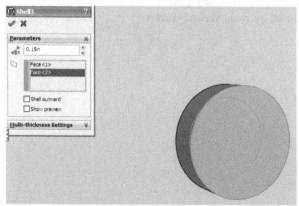

Figure 6-32 Preview for shelling operation

Figure 6-33 Completed part

Cap with internal thread

18. Select the narrow back face. Click the Hidden Lines Removed option.
19. Click Feature > Reference Geometry > Plane.
20. Enter 0.45 in for the distance.
21. Click the Reverse direction box.
22. Click OK from the PropertyManager of the plane. Plane1 is displayed in the FeatureManager (see Fig. 6-34).
23. Right-click Plane1 from the FeatureManager.
24. Click Sketch from the toolbar.
25. Click the inner face of the shell.
26. Click the Convert Entities Sketch tool. The inner face is displayed on Plane1 (see Fig. 6-35).
27. Click on Features > Curves > Helix and Spiral (or Insert > Curves > Helix and Spiral). The Helix/Spiral dialog appears.
28. Define the helix by Pitch (0.125 in) and Revolution (3) and check Clockwise (see Fig. 6-36).

29. Create a plane, Plane1, normal to the end of the spring by clicking the endpoint of the spring and clicking Feature > Reference Geometry > Plane. Alternatively, click Sketch on the right-hand plane. The Helix PropertyManager appears. On FeatureManager, click on the feature, Helix/Spirial1, with this endpoint, <Point1>.

30. Click on Plane1 and create a circle of diameter 0.125 in, with center at the origin of the helix.

31. Exit sketch mode.

32. Click the Swept Boss/Base tool. The Sweep PropertyManager appears.

33. Right-click the Profile and Path box.

34. Click the circle as the profile and the helix as the path. It is best to do this using the FeatureManager. A real-time preview will appear.

35. Click OK to complete the swept cut part, which is hollow (see Fig. 6-37). Hide the planes.

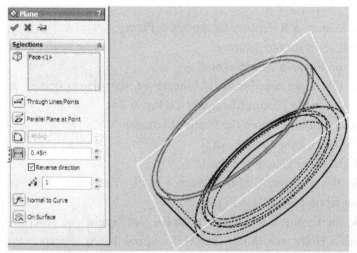

Figure 6-34 Creating Plane1

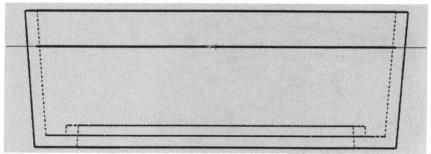

Figure 6-35 Extracting the inner face of the shell using the Convert Entities Sketch tool

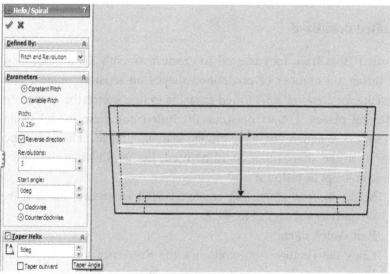

Figure 6-36 Preview of the helix for internal threading

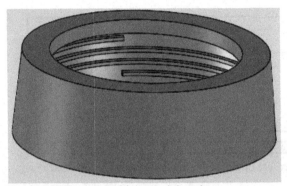

Figure 6-37 Final part with internal thread

6.3 Lofted Boss/Base

The Lofted Boss/Base tool is used to create a smooth 3D surface that passes through a number of predefined shapes on separate planes. The prerequisite for creating a 3D lofted model is to first sketch the shapes on the different planes. Let us illustrate the lofted boss/base model using three shapes: a square base, a circle, and an ellipse. There are no restrictions in the shapes that could be used. Let us illustrate the lofted boss/base concept as follows:

1. Select the top plane and create Sketch1 (a square, 90 mm by 90 mm) as shown in Fig. 6-38.
2. Exit sketch mode.
3. Click the Features tool and click the Reference tool. Select the Plane option. The plane box appears; select the top plane.
4. Set the distance between the existing top plane and a new reference plane as 50 mm. Click OK checkbox. A new plane, Plane1, appears.
5. Create Sketch2 (a circle of diameter 60 mm), as shown in Fig. 6-40.
6. Exit sketch mode.
7. Click the Features tool and click the Reference tool. Select the Plane option. The plane box appears; select the top plane.

8. Set the distance between Plane1 and a new reference plane as 50 mm. Click the OK checkbox. A third plane, Plane2 appears, as shown in Fig. 6-41.
9. Create Sketch3 (an ellipse of major diameter 80 mm and minor diameter 65 mm), as shown in Fig. 6-40.
10. Exit sketch mode. All sketches appear as in Fig. 6-41.
11. Click the Lofted Boss/Base tool. The Loft PropertyManager appears.
12. Right-click the Profiles box.
13. Click the square, then the circle, and finally the ellipse. A real-time preview will appear (see Fig. 6-42).
14. Click OK to complete the lofted part (see Fig. 6-43). Hide the planes.

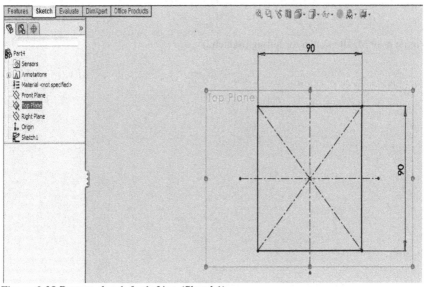

Figure 6-38 Bottom sketch for lofting (Sketch1)

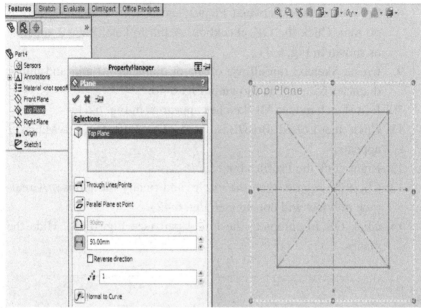

Figure 6-39 Middle sketch for lofting (Sketch2)

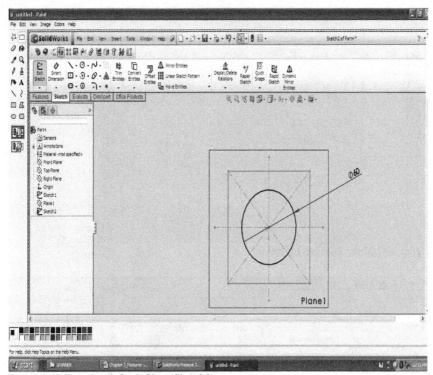

Figure 6-40 Top sketch for lofting (Sketch3)

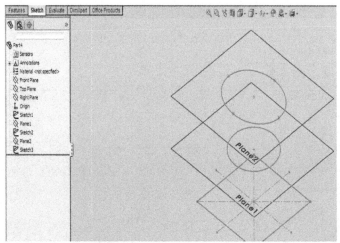

Figure 6-41 Three sketches for lofting

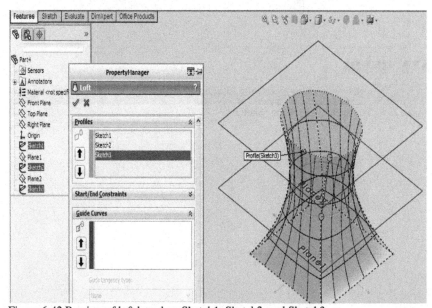

Figure 6-42 Preview of loft based on Sketch1, Sketch2, and Sketch3

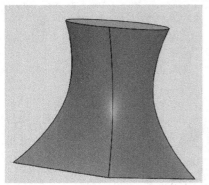

Figure 6-43 Final 3D lofted model

6.3.1. *Practical examples*

This section gives two practical examples of lofted parts. The first example is an impeller, commonly used in compressors and turbines, and the second is an aircraft wing.

Impeller

1. Select the front plane and create Sketch1 (a circle of diameter 3 in) and extrude by 0.08 in (see Fig. 6-44).
2. Create Sketch2 (a circle of diameter 0.6 in) and extrude by 1.5 in (see Fig. 6-45).

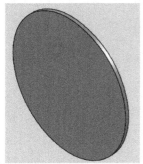

Figure 6-44 Extrude1 from Sketch1

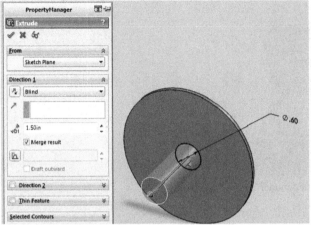

Figure 6-45 Extrude2 from Sketch2

3. Create Sketch3 (two arcs, with radii 1.05 in and 1.03, in respectively) on top of the first extruded face. The center of the first arc is (0.75 in, 0.81 in) away from the center of Sketch1; the center of the other is offset 0.08 in from the first, as shown in Fig. 6-46. Use Convert Entities to extract the concentric circles (diameters 3 in and 0.6 in, respectively). Trim the circles from Sketch3 (see Fig. 6-47).

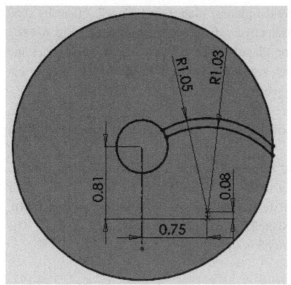

Figure 6-46 Geometric definitions for fin profile on top of Extrude1

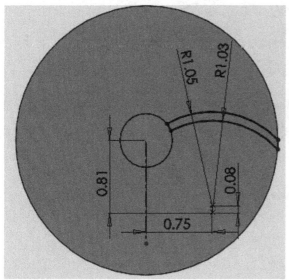

Figure 6-47 Trimmed shapes which fully define the fin profile on top of Extrude1

4.   Extrude Sketch3 to 0.6 in (see Fig. 6-48).

5. Define two reference planes, Plane1 and Plane2, offset by 0.68 in and 0.85 in, respectively, from the front plane (see Fig. 6-48).

6. Create Sketch4 on Plane1, select all edges of the fin, extract and convert to entities using the Convert Entities tool. Notice that Plane1 is active (see Fig. 6-49).

7. Exit sketch mode.

8. Create Sketch5 on Plane2. Create a circle of diameter 1.77 in (see Fig. 6-50).

9. On Sketch5, add two arcs so that their ends touch the circles. Both arcs have a radius of 0.62 in, and an angle of 30°; their centers are offset from one another by 0.08 in (see Fig. 6-50). The geometry, Sketch5 is trimmed as shown in Fig. 6-51. The fully defined Sketch5 is shown in Fig. 6-52.

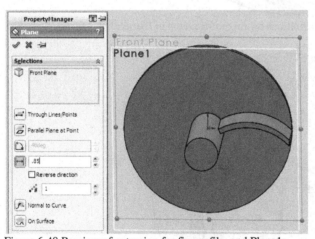

Figure 6-48 Preview of extrusion for fin profile, and Plane1

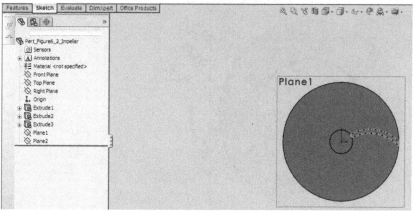

Figure 6-49 Extracting the fin profile onto Plane1 from Extrude2

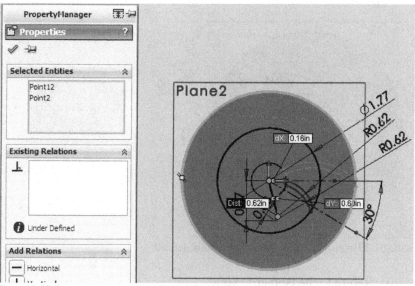

Figure 6-50 Geometric definitions for the fin profile on Plane2

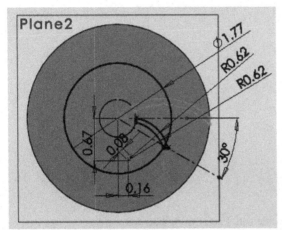

Figure 6-51 Geometries to be trimmed for fin profile on Plane2

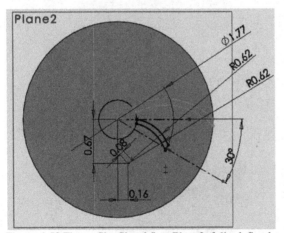

Figure 6-52 Fin profile, Sketch5 on Plane2, fully defined

10. Return to the isometric view.
11. Click the Features tool and click the Reference tool. Select the Plane through Vertices option. Create two planes, Plane3 (defined using the left-hand vertices of Sketch4 and Sketch5 and one other point) and Plane4 (defined using the right-hand vertices of Sketch4 and Sketch5 and one other point).

12. Sketch a spline using the 3D Sketch Tool on Plane3 to define a lofting profile, Profile1. Sketch another spline for a second lofting profile, Profile2, on Plane4, as shown in Fig. 6-53.

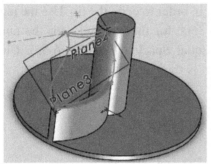

Figure 6-53 Creating Plane3 and Plane4

13. Create a loft base using Sketch4 and Sketch5 as profiles, Profile1 and Profile2 as guide curves, as shown in Fig. 6-54.

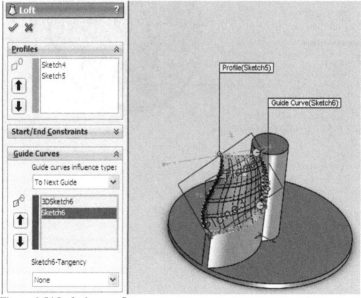

Figure 6-54 Lofted upper fin

14. Click OK and hide Plane3 and Plane4.
15. Click View > Temporary Axes. This displays the axes. Look for the axis at the top of the cylinder; it may be very small and not easy to spot.
16. Click Circular Pattern in order to copy the fin. Use the temporary axis, Axis<2>, enter 360° for the angle, 12 for the number of patterns, and check Equal Spacing (see Fig. 6-55). Click OK. The fully modeled part is shown in Fig. 6-56

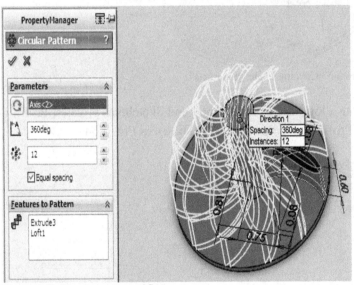

Figure 6-55 Circular pattern of fins

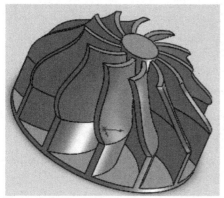

Figure 6-56 Final impeller

Aircraft wing

1. Select the right-hand plane and create the polygon in Sketch1, as shown in Fig. 6-57.
2. Use the Spline Tool with the polygon to sketch a spline profile.
3. Create Plane1 10 in from the right-hand plane (see Fig. 6-58).
4. On Plane1, select all the edges of Sketch1 and convert them to entities using the Convert Entities tool. This ensures that the two profiles, on the right-hand plane and Plane1, are the same.
5. Copy the converted entities so that they start 4 in away from the origin. Note that the overall length is 6 in (see Fig. 6-59).
6. For the copied profile, use the Smart Dimension Tool to change the length from 6 in to 3.5 in (see Fig. 6-60). Delete the larger profile on Plane1, and retain the smaller profile, as in Fig. 6-61.
7. Exit sketch mode.
8. Create a loft using the two remaining profiles (see Fig. 6-62 for the profiles, Fig. 6-63 for the preview, and Fig. 6-64 for the final part).
9. Hide Plane1.

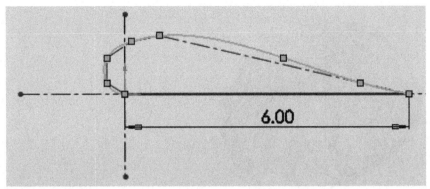

Figure 6-57 Sketch1

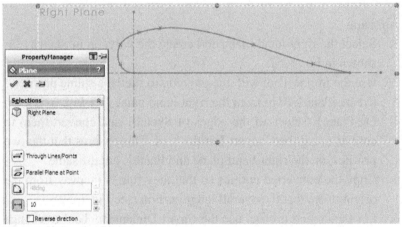

Figure 6-58 Creating Plane1

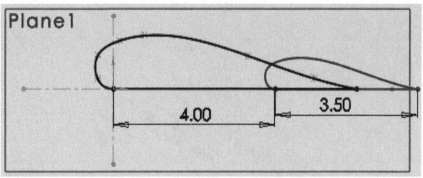

Figure 6-59 Convert entities on the right-hand plane to entities on Plane1, and copy entities to 4 in away

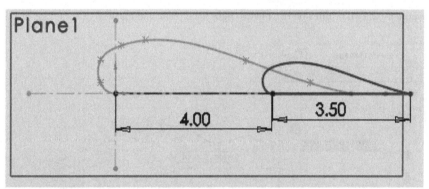

Figure 6-60 Change size of copied profile by resizing 6 in to 3.5 in

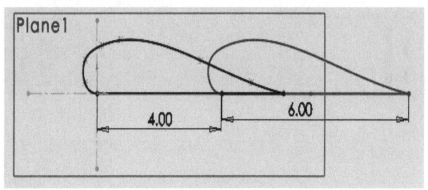

Figure 6-61 Delete the larger profile and retain the resized profile, which has a length of 3.5 in and is 4 in from the origin

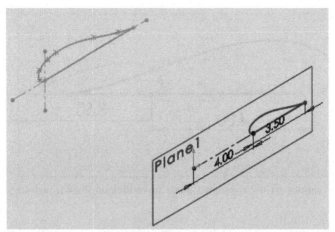

Figure 6-62 Two profiles on two planes (right-hand plane and Plane1)

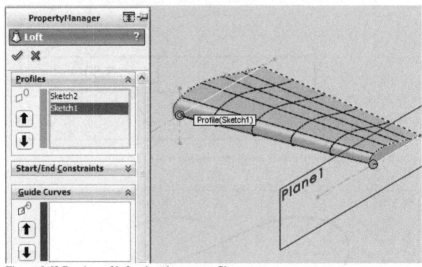

Figure 6-63 Preview of loft using the two profiles

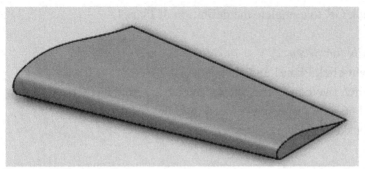

Figure 6-64 Final lofted wing

6.4 Practical Swept Feature: Elbow Casting for Steam Valve

Sketch path for sweep
1. Open a new part document.
2. Select the front plane.
3. Sketch a quarter-circle, Sketch1, with radius equal to 70 mm (see Fig. 6-65).
4. Exit sketch mode.

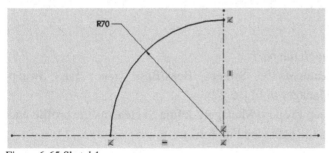

Figure 6-65 Sketch1

Create a plane normal to the endpoint of Sketch1
5. Click the left endpoint of Sketch1.
6. Select the reference geometry icon and click Features > Plane.
7. In the Section rollout, select Sketch1.
8. Select Normal To Curve.

9. Click OK to complete the definition of Plane1.

Sketch profile for sweep

10. Right-click Plane1 and click Normal To.
11. Sketch two concentric circles, Sketch2, with diameters of 82 mm and 66 mm, respectively (see Fig. 6-66).
12. Exit sketch mode.

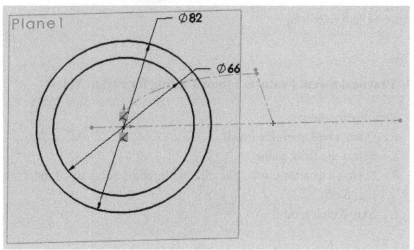

Figure 6-66 Sketch2

Sweep profile through the part

13. Click Features > Sweep Boss/Base (see the Sweep PropertyManager in Fig. 6-67).
14. In the Sweep PropertyManager, define Sketch2 as the profile and Sketch1 as the path (see Fig. 6-67).

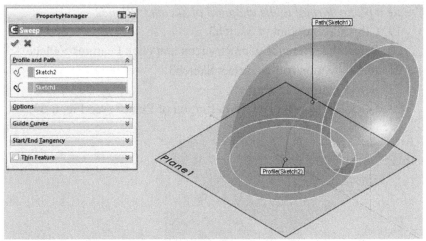

Figure 6-67 Sweep PropertyManager

Sketch a line at 45° to the horizontal. (This line will be used to create a plane normal to its endpoint.)

15. Click the front plane.
16. Click Sketch and Normal To.
17. Sketch a dimension line, Sketch3, 152.7 mm long, from the origin at an angle of 45°, and dimension it from the origin (152.7 = (124 - 16)/Cosine(45°)), as shown in Fig. 6-68.
18. Exit sketch mode.

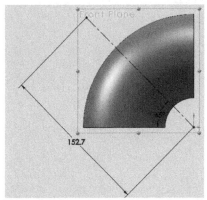

Figure 6-68 Line sketched for creating a plane

Create a plane normal to the endpoint of Sketch3: Plane2

19. Click the endpoint of Sketch3.
20. Select the reference geometry icon and click Features > Plane.
21. In the Section rollout, select Sketch3.
22. Select Normal To Curve.
23. Click OK to complete the definition of Plane2 (see Fig. 6-69).

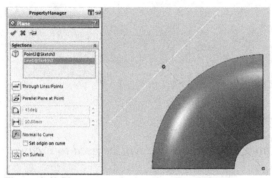

Figure 6-69 Plane created normal to line defined

Create Sketch4 on Plane2

24. Click Plane2 and rotate appropriately, but do not choose Normal To.
25. Create two circles, Sketch4, with diameters of 45 mm and 58 mm, respectively (see Fig. 6-70).

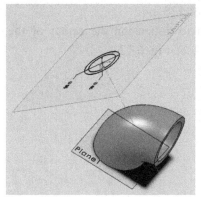

Figure 6-70 Concentric circles sketched on Plane2

Extrusion of concentric circles

26. Click Extrude.

27. In the Extrude PropertyManager, click Up Too Surface for Direction 1, and select the inner surface of the elbow (see Fig. 6-71).

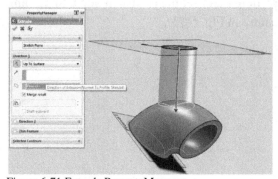

Figure 6-71 Extrude PropertyManager

Create right-hand flange for elbow

28. Click the right-hand face of the elbow and click Sketch, to start sketch mode.

29. Click Normal To; this makes the face selected normal to the viewer.

30. Sketch a circle with a diameter of 128 mm.

31. Use the Smart Dimension tool to dimension the center of the circle at 70 mm from the elbow (see Fig. 6-72).

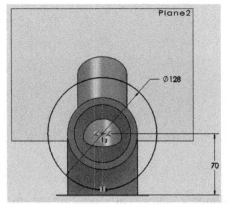

Figure 6-72 Circular profile for right-hand elbow flange

32. Click the Extrude tool from CommandManager.

33. Click Blind for Direction 1 and set the extrusion depth to 12 mm (see the extruded feature in Fig. 6-73).

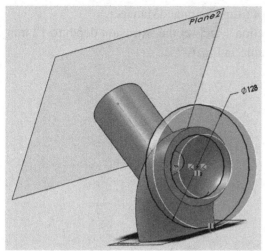

Figure 6-73 Extrusion to complete right-hand elbow flange

Create bottom flange for elbow

34. Click Plane1 at the bottom face of the elbow, and click Sketch to start sketch mode.

35. Click Normal To; this makes the face selected normal to the viewer.

36. Sketch a circle with a diameter of 125 mm (see Fig. 6-74).

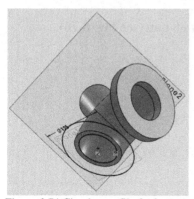

Figure 6-74 Circular profile for bottom elbow flange

37. Click the Extrude tool from CommandManager.
38. Click Blind for Direction 1 and set the extrusion depth to 12 mm (see the extruded feature in Fig. 6-75).

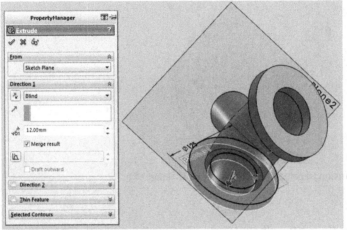

Figure 6-75 Extrusion to complete bottom elbow flange

Create top flange for elbow

39. Click Plane2 at the top face of the elbow and click Sketch to start sketch mode
40. Click Normal To; this makes the face selected normal to the viewer.
41. Sketch a circle with a diameter of 88 mm (see Fig. 6-76).

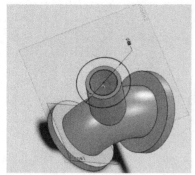

Figure 6-76 Circular profile for top elbow flange

42. Click the Extrude tool from CommandManager.
43. Click Blind for Direction 1 and set the extrusion depth to 12 mm (see the extruded feature in Fig. 6-77). The final elbow model is shown in Fig. 6-78.

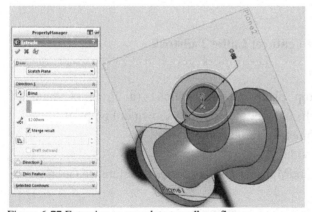

Figure 6-77 Extrusion to complete top elbow flange

Figure 6-78 Final elbow model

6.5   Practical Swept Feature: Lathe Tailstock

Tailstock ring
1. Select the top plane. Create Sketch1 (a circle of diameter 176 mm) with center at the origin as shown in Fig. 6-79.
2. Select the isometric orientation.
3. Exit sketch mode.
4. From CommandManager click Features > Reference Geometry > Plane.
5. Select the Normal to Curve option.
6. Click Sketch1.
7. Click OK to create Plane1, normal to Sketch1.
8. Right-click Plane1 and create Sketch2 (a circle of diameter 22 mm)
9. Add a piercing relation by clicking the center of the small circle, then hold the Ctrl key down. Click somewhere on the

circumference of the large circle and release the Ctrl key. Right-click Make Pierce from the dialog. Click OK (see Fig. 6-80).

10. Exit sketch mode.
11. Click the Swept Boss/Base tool. The Sweep PropertyManager appears.
12. Right-click the Profile and Path box.
13. Click the small circle (Sketch2) as the profile and the large circle (Sketch1) as the path using FeatureManager. A real-time preview will appear, as shown in Fig. 6-81.
14. Click OK to complete the swept cut part, which is hollow as shown in Fig. 6-81. Hide the planes.

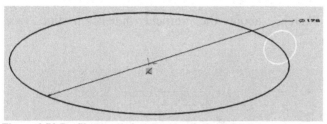

Figure 6-79 Profile (small circle) and path (large circle)

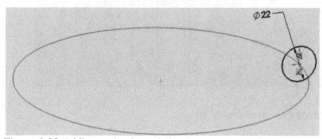

Figure 6-80 Adding a piercing relation

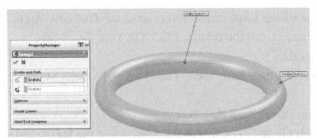

Figure 6-81 Sweep PropertyManager and preview of tailstock ring

Tailstock central boss

15. Select the right-hand plane.
16. Create Sketch3, 8 mm offset from the origin, as shown in Fig. 6-82.
17. Revolve Sketch3 about the vertical dimension line (see Fig. 6-83).

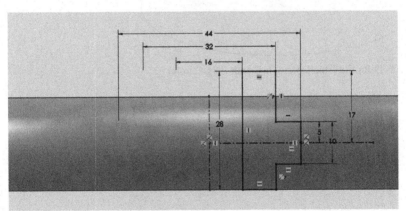

Figure 6-82 Sketch2 to create revolved central boss

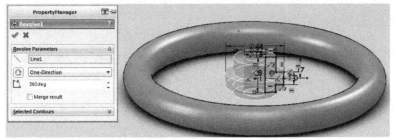

Figure 6-83 PropertyManager and preview of the revolved central boss

Create two planes

18. Select the right-hand plane.
19. Select the reference geometry icon and click Features > Plane.
20. In the Distance rollout, set a value of 18 mm (see Fig. 6-84).
21. Click OK to complete the definition of the plane.
22. Select the right-hand plane.
23. Select the reference geometry icon and click Features > Plane.
24. In the Distance rollout, set a value of 82 mm (see Fig. 6-85).
25. Click OK to complete the definition of the plane.

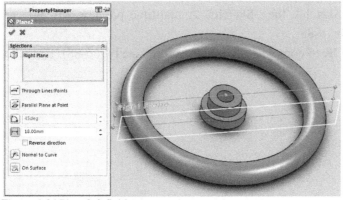

Figure 6-84 Plane2 definition

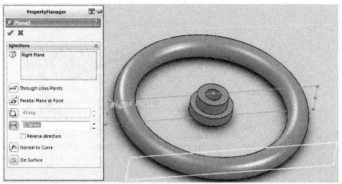

Figure 6-85 Plane3 definition

Lofting

26. Select Plane2, and create a rectangular profile, Sketch4 (see Fig. 6-86).
27. Select Plane3, and create a rectangular profile, Sketch5 (see Fig. 6-87).
28. Click the Lofted Boss/Base tool. The Loft PropertyManager appears.
29. Right-click the Profiles box.
30. Click Sketch4, then Sketch5. A real-time preview will appear (see Fig. 6-88).
31. Click OK to complete the lofted feature (see Fig. 6-88). Hide the planes.

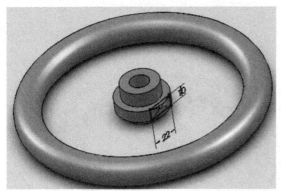

Figure 6-86 Sketch4 on Plane2

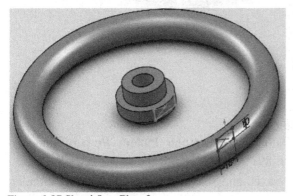

Figure 6-87 Sketch5 on Plane3

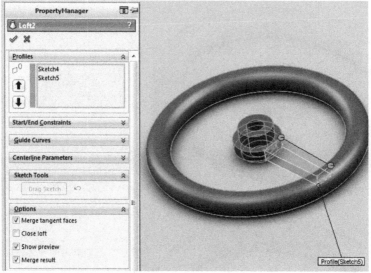

Figure 6-88 Preview of lofted rib

Create circular pattern of ribs

32. Click Circular Pattern, the Circular Pattern PropertyManager appears (see Fig. 6-89).
33. Set the temporary axis, and select it as the axis for the circular pattern.
34. Set the number of instances to 4.
35. Select Loft2 as the Features to Pattern.
36. Add fillets.
37. Click OK to complete the design, as shown in Fig. 6-90.

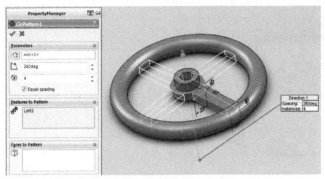

Figure 6-89 Circular Pattern PropertyManager

Figure 6-90 Final model of lathe tailstock

Exercise

Repeat the design of the tail stock ring of Section 6.5. Create an ellipse with a major diameter of 22 mm and a minor diameter of 10 mm. Create another ellipse with a major diameter of 18 mm and a minor diameter of 10 mm. Loft between these features and use the Circular Pattern tool to replicate the lofted feature four times. Follow the steps shown in Figs 6-91 to 6-94.

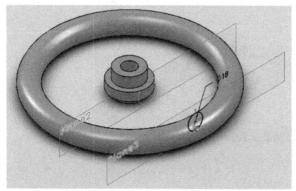

Figure 6-91 Ellipses

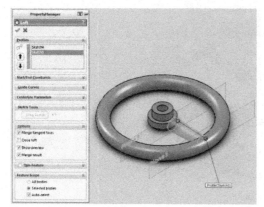

Figure 6-92 Preview of loft

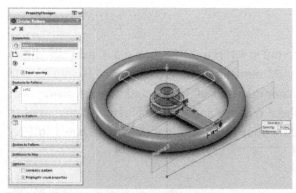

Figure 6-93 Preview of circular pattern

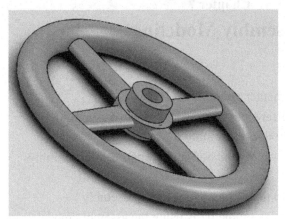

Figure 6-94 Final tailstock

Chapter 7

Assembly Modeling

Objectives:
When you complete this chapter you will:
- Understand the differences between top-down assembly modeling and bottom-up assembly modeling
- Have learnt how to apply the bottom-up approach to assembly modeling
- Have learnt how to mate components in an assembly
- Have learnt how to assemble parts using the assembly model methodology
- Have learnt how to use assembly analysis (or interference analysis) to analyze how good an assembly is, and also to get a feel of the quality of the parts design
- Have learnt how to use the Exploded View Assembly Tool
- Have learnt how to animate an exploded view

7.1 Introduction

Assembly modeling is the combining of part models into complex, interconnected solid models. There are two well-known approaches to assembly modeling: top-down and bottom-up.

Top-down assembly modeling
In top-down assembly modeling, major design requirements are translated into assemblies, sub-assemblies, and components.

Bottom-up assembly modeling
In bottom-up assembly modeling, components are developed independently, based on the design requirements, and combined into sub-assemblies and assemblies. The three basic steps in bottom-up assembly modeling are as follows:

- Create each component independent of any other component in the assembly;
- Insert the components into the assembly; and
- Mate the components in the assembly according to the physical constraints of the design.

The bottom-up approach is used in this book. Since mating the components in an assembly is a new concept, this is first presented and then the entire process of bottom-up assembly modeling is applied to typical assembly problems.

7.2   Starting the Assembly Mode of SolidWorks

There are two ways to start the assembly mode of SolidWorks: from a new SolidWorks document or from an existing part that we wish to place as the first one in the assembly.

1. To start the assembly mode of SolidWorks, invoke the New SolidWorks Document dialog, choose the Assembly button and click OK, as shown in Fig. 7-1. The Begin Assembly PropertyManager pops up, as shown in Fig. 7-2. Browse to open the first part to add.

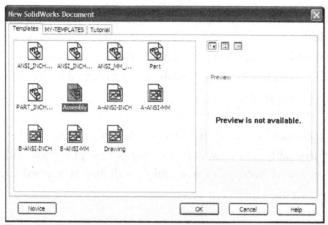

Figure 7-1 The New SolidWorks Document dialog

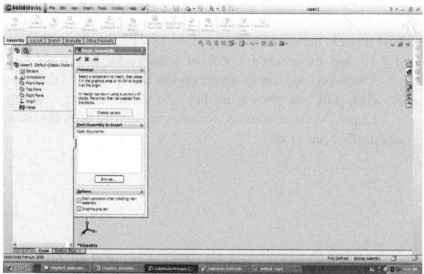

Figure 7-2 Assembly mode showing the Begin Assembly PropertyManager

2. From an existing part document, click the New dialog as shown in Fig. 7-3. Click Make Assembly from Part/Assembly. The

Begin Assembly PropertyManager pops up, as shown in Fig. 7-2. Browse to open the first part to add.

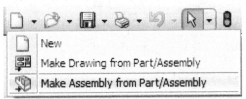

Figure 7-3 Accessing Assembly mode from the New dialog in an existing part document

7.3 Inserting Components into an Assembly Document

There are several ways of inserting component into an assembly document in SolidWorks:

1. Click Assembly > Insert Components (see Fig. 7-4).

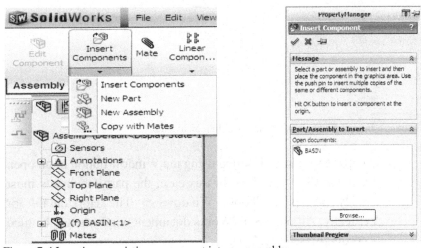

Figure 7-4 Inserting an existing component into an assembly

2. Click Insert > Component > Existing Part/Assembly (see Fig. 7-5).

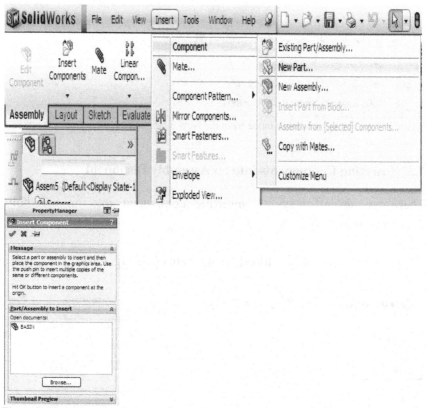

Figure 7-5 Inserting an existing component into an assembly

3. Components can be inserted using the Window option of an open document (See Fig. 7-6). In this case, the part documents must already be open. Choose Windows > Tile Horizontally or Vertically. All open SolidWorks document windows will be tiled accordingly (see Fig. 7-7).

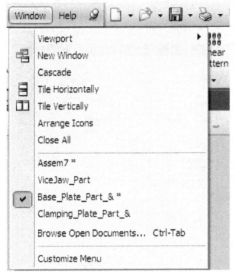

Figure 7-6 Using the Window option of an open document

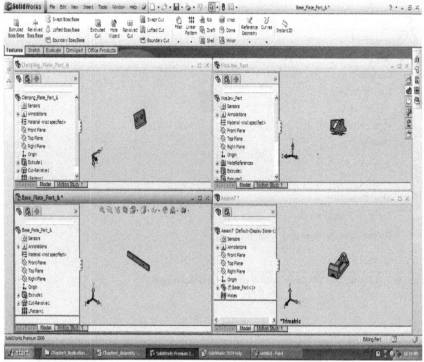

Figure 7-7 Tiled windows

7.4 Mates

A mate is a geometric relationship between assembly components. There are three types of mate in SolidWorks: standard, advanced, and mechanical. See Fig. 7-8. Standard mates create geometric relationships such as Coincident, Parallel, Perpendicular, Tangent, Concentric, Lock, Distance, and Angle. Each mate is valid for specific combinations of geometry. A comprehensive list of possible mates for different types of part is given here.

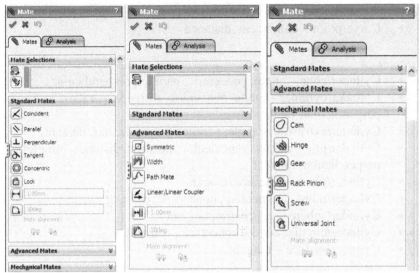

Figure 7-8 Standard, advanced, and mechanical mates in SolidWorks

Circular or arc edge

- Circular or arc edge/cone – coincident, concentric
- Circular or arc edge/line – concentric
- Circular or arc edge/cylinder – concentric, coincident
- Circular or arc edge/plane – coincident
- Circular or arc edge/circular or arc edge – concentric

Cone

- Cone/circular or arc edge – coincident, concentric
- Cone/cone – angle, coincident, concentric, distance, parallel, perpendicular
- Cone/cylinder – angle, concentric, parallel, perpendicular
- Cone/extrusion – tangent
- Cone/line – angle, concentric, parallel, perpendicular
- Cone/plane – tangent
- Cone/point – coincident, concentric
- Cone/sphere – tangent

Curve
- Curve/point – coincident, distance

Cylinder
- Cylinder/cone – angle, concentric, parallel, perpendicular
- Cylinder/cylinder – angle, concentric, distance, parallel, perpendicular, tangent
- Cylinder/extrusion – angle, parallel, perpendicular, tangent
- Cylinder/line – angle, coincident, concentric, distance, parallel, perpendicular, tangent
- Cylinder/plane – distance, tangent
- Cylinder/point – coincident, concentric, distance
- Cylinder/sphere – concentric, tangent
- Cylinder/circular edge – concentric, coincident
- Cylinder/surface – tangent

Extrusion
- Extrusion/cone – angle, parallel, perpendicular
- Extrusion/cylinder – angle, parallel, perpendicular, tangent
- Extrusion/extrusion – angle, parallel, perpendicular
- Extrusion/line – angle, parallel, perpendicular
- Extrusion/plane – tangent
- Extrusion/point – coincident

Line
- Line/cone – angle, concentric, parallel, perpendicular
- Line/cylinder – angle, coincident, concentric, distance, parallel, perpendicular, tangent
- Line/extrusion – angle, parallel, perpendicular
- Line/line – angle, coincident, distance, parallel, perpendicular
- Line/plane – coincident, distance, parallel, perpendicular
- Line/point – coincident, distance
- Line/sphere – concentric, distance, tangent
- Line/circular edge – concentric

Plane
- Plane/cone – tangent
- Plane/cylinder – distance, tangent
- Plane/extrusion – tangent

- Plane/line – coincident, distance, parallel, perpendicular
- Plane/plane – angle, coincident, distance, parallel, perpendicular
- Plane/point – coincident, distance
- Plane/sphere – distance, tangent
- Plane/circular edge – coincident
- Plane/surface – tangent

Point

- Point/cone – coincident, concentric
- Point/curve – coincident, distance
- Point/cylinder – coincident, concentric, distance
- Point/extrusion – coincident
- Point/line – coincident, distance
- Point/plane – coincident, distance
- Point/point – coincident, distance
- Point/sphere – coincident, concentric, distance
- Point/surface – coincident

Sphere

- Sphere/cone – tangent
- Sphere/cylinder – concentric, tangent
- Sphere/line – concentric, distance, tangent
- Sphere/plane – distance, tangent
- Sphere/point – coincident, concentric, distance
- Sphere/sphere – concentric, distance, tangent

Surface

- Surface/cylinder – tangent
- Surface/plane – tangent
- Surface/point – coincident

7.5   Assembly Modeling Methodology

Initially, two windows should be open: one for the fixed part and another for the assembly.

A. Open a window for the fixed part

> Click Open and browse for the path where the fixed part is located
> Select Part for Files of Type, for example, Part*.sldprt
> Click View Menu > check Thumbnails
> Double-click Part of interest to be fixed, the Part FeatureManager is displayed

B. Open a window for the assembly

> Click New
> Double-click Assembly from the default Templates tab
>
> Double-click Part in the Open documents box
> Click OK from the Begin Assembly PropertyManager
> Click Window > Part > File > Close   (Close the window for the fixed part and leave only the assembly window open.)

C. Set assembly units

> Click Options > Documents > Units > MMGS (IPS) > OK > Save (to set units for the assembly)

D. Assemble parts

For each part to be inserted into the assembly:
- Click Assembly > Insert Components > Browse for existing parts
- Double-click a part from the path as Part*.sldprt format
- Click a location to position the part
- Click the Mate Assembly Tool and check Faces to Mate

Next part

E. Exploded assembly view tool

For each component to explode:
- Click the component
- Drag the blue manipulator handle in the appropriate direction
- Click Done

Next part

F. Animate the exploded view

Right-click ExplView1 from the ConfigurationManager
Click Animate Explode
Click Stop
Close the Animator Controller
Right-click Collapse

G. Exploded assembly view tool

Right-click the Component to view in the graphics window
Click Open Part from the shortcut toolbar
Click front plane (or any other plane)
Click Section View for the Heads-Up View Toolbar

7.6 Creating the Components

The first task in assembly modeling is to create the various components or parts that constitute the assembly. The files containing these parts should be saved in an appropriate directory of the computer being used. The bench vise assembly project is used to illustrate the assembly process.

Jaw screw
1. Sketch1 is half of the overall sketch. See Fig. 7-9.
2. Revolve Sketch1 about the horizontal centerline. See Fig. 7-10. The partly completed jaw screw part is shown in Fig. 7-11.

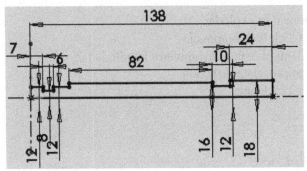

Figure 7-9 Sketch1

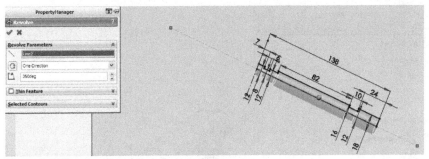

Figure 7-10 Revolve Sketch1

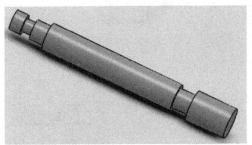

Figure 7-11 Partly completed jaw screw part

Inserting a hole

3. Define Plane1. See Fig. 7-12.
4. Extrude-cut a 6 mm diameter hole on Plane1, all the way through the head of the jaw screw. See Fig. 7-13.
5. Hide Plane1. Fig. 7-14 shows the completed jaw screw.

Figure 7-12 Plane1 defined

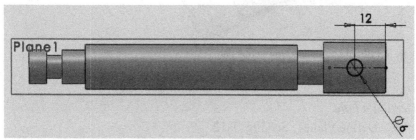

Figure 7-13 Extrude-cut a 6 mm diameter hole on Plane1

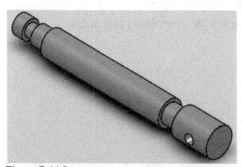

Figure 7-14 Jaw screw

Screw Bar

1. Sketch1 is half of the overall sketch. See Fig. 7-15.
2. Revolve Sketch1 about the horizontal centerline. See Fig. 7-16. The screw bar is shown in Fig. 7-17.

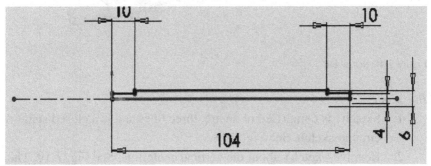

Figure 7-15 Sketch1

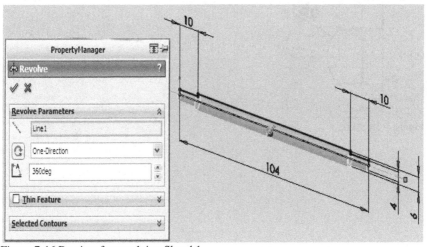

Figure 7-16 Preview for revolving Sketch1

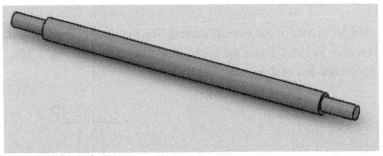

Figure 7-17 Screw bar

Bar globe

1. Sketch1 is comprised of an arc, three lines and is a closed semi-circular sketch. See Fig. 7-18.
2. Revolve Sketch1 about the vertical centerline. See Fig. 7-19. The revolved sketch is shown in Fig. 7-20.

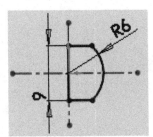

Figure 7-18 Sketch1

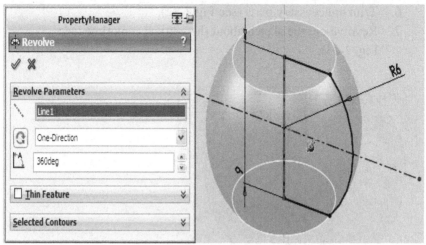

Figure 7-19 Preview for revolving Sketch1

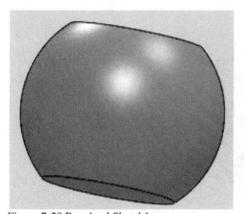

Figure 7-20 Revolved Sketch1

3. Choose the front plane.
4. Use the Convert Entities tool to extract the top and bottom circular entities, which become straight lines on the front plane as shown in Fig 7-21.
5. Sketch a slanting line at 45° to the horizontal.

6. Trim unnecessary parts (see Fig 7-22).
7. Revolve-cut the sketch about the vertical centerline (see Fig. 7-23).

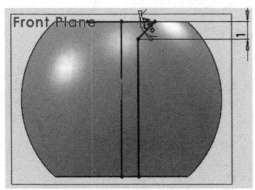

Figure 7-21 Convert Entities tool used on the front plane to extract features

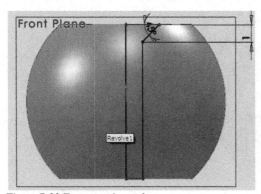

Figure 7-22 Features trimmed

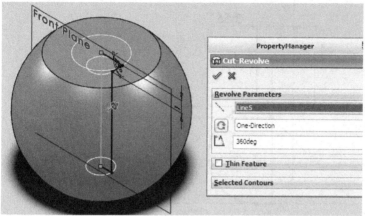

Figure 7-23 Preview of bar globe

Alternative method of modeling a bar globe

Using geometry, we can calculate that if the diameter of the arc used in Sketch1 is 12 mm, which is a truncated circle so that the height of the bar globe is 9 mm, then the length of the flat sections at the top and bottom of the globe is 3.558 mm (3.56 mm). See Fig. 7-24, which gives a new Sketch1 for the bar globe.

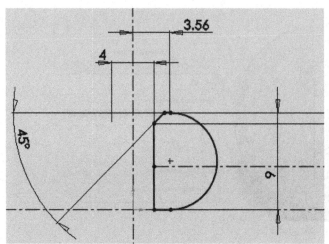

Figure 7-24 Sketch1 for alternative method of modeling a bar globe

1. Revolve Sketch1 about the vertical construction line. See Figs 7-25 and 7-26.

This shows that there often several ways to model a part. The second approach is much more efficient in terms of modeling time.

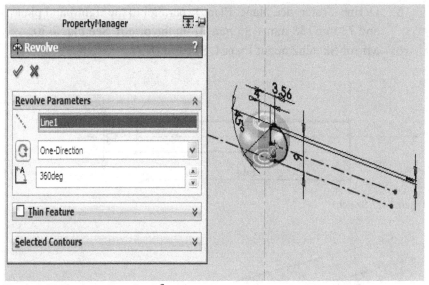

Figure 7-25 Preview of revolved feature

Figure 7-26 Bar globe

Base plate

1. Sketch the base (Sketch1) symmetrically on the top plane. See Fig. 7-27.
2. Extrude by 6 mm. See Fig. 7-28.
3. Select the front plane and sketch a small 5-sided shape (screw hole, for a 6 mm diameter screw, V-shaped at the top), Sketch2, for revolving. See Fig. 7-29.
4. Revolve-cut Sketch2 about the vertical construction line. See Fig. 7-30.

5. Define a reference plane, Plane1, parallel to the right-hand plane and 73 mm (38 mm + 35 mm) from the origin. See Fig. 7-30.
6. Mirror Sketch2 about Plane1. See Fig. 7-31.

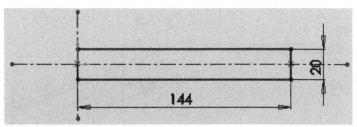

Figure 7-27 Sketch1

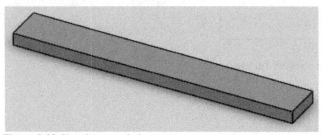

Figure 7-28 Sketch1 extruded

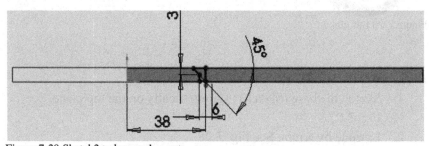

Figure 7-29 Sketch2 to be revolve-cut

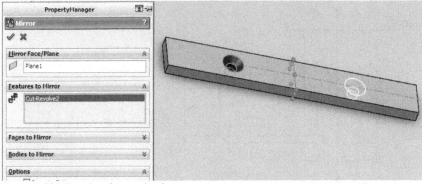

Figure 7-30 Mirroring the revolved cut

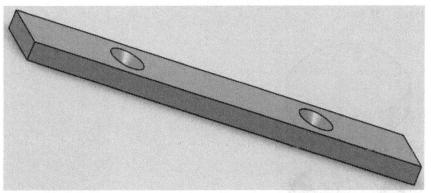

Figure 7-31 Base plate

Oval fillister

1. Choose the front plane and position Sketch1. First, sketch the lower part, mirror it, and then sketch the arc of radius 6 mm. Remove the mirrored half and trim the arc. See Fig. 7-32.

2. Revolve Sketch1 about the vertical construction line. See Fig. 7-33.

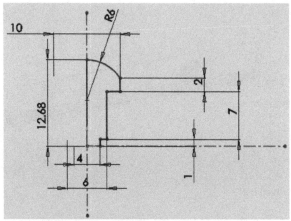

Figure 7-32 Sketch1

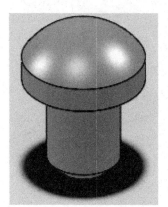

Figure 7-33 Revolved partially completed part

3. Activate the right-hand plane.
4. Sketch a rectangle on the right-hand plane, 2 mm by 2 mm, positioned slightly above the top of the oval fillister. See Fig. 7-34.
5. Extrude-cut in the Mid-Plane by 15 mm, to completely cut right through. See Fig. 7-35.

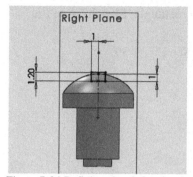

Figure 7-34 Defining sketch for slot

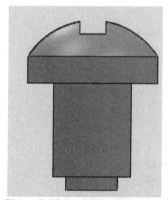

Figure 7-35 Oval fillister

Set screw 1

This set screw will be created manually, and will not have a thread. Threaded screws are created using the Design Library Toolbox, described in Chapter 19.

1. Choose the front plane.
2. Sketch half of the profile, as Sketch1. See Fig. 7-36.
3. Revolve Sketch1 about the vertical construction line. See Fig. 7-37.

4. Choose the top plane and sketch a rectangle, Sketch2. See Fig. 7-38.

5. Extrude-cut by 2 mm downward. See Fig. 7-39.

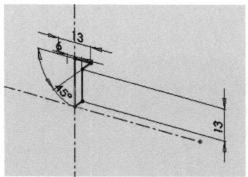

Figure 7-36 Sketch1

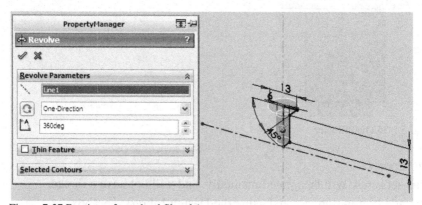

Figure 7-37 Preview of revolved Sketch1

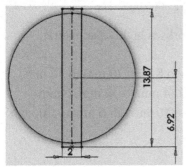

Figure 7-38 Sketch for slot

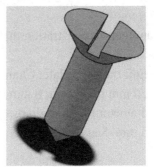

Figure 7-39 Set screw 1

Set screw 2

SolidWorks is highly parametric; the model of the second set screw is produced by changing that of the first.

1. Edit the first set screw, with no angular dimension, and change the screw head height to 5 mm and the length of the screw to 18 mm. See Fig. 7-40.

Figure 7-40 Set screw 2

Clamping plate

This is another demonstration that SolidWorks is highly parametric.

1. Edit the base plate. Change 38 mm to 8 mm, 70 mm to 16 mm, 20 mm to 32 mm, 6 mm to 7 mm, 144 mm to 32 mm, 6 mm diameter to 7 mm diameter, 45° to 60°, through 6 mm to through 5 mm. See Fig. 7-41.

Figure 7-41 Clamping Plate

Vise jaw

1. Select the right-hand plane.
2. Sketch 5 lines (2 vertical and 3 horizontal) on the right-hand side of the vertical centerline.
3. Mirror these lines about the vertical centerline.
4. Sketch an arc using the 3-Point Arc tool and use a relation to center the arc about the vertical centerline. Dimension the arc to have a radius of 16 mm.
5. Sketch two slanting lines.

6. Apply relations: the slanting lines are tangential to the arc, they end at the horizontal lines and the highest point of the arc is on the vertical centerline.
7. Sketch1 is complete. See Figs 7-42 and 7-43.
8. Extrude Sketch1 through 22 mm (32 mm – 10 mm). See Fig. 7-44.

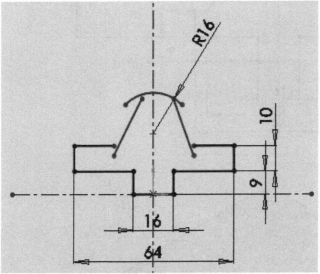

Figure 7-42 Elements of Sketch1

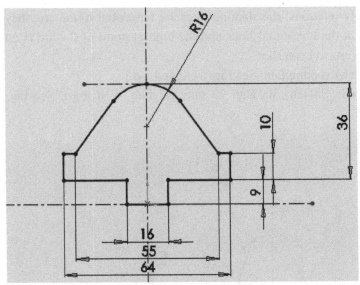

Figure 7-43 Sketch1

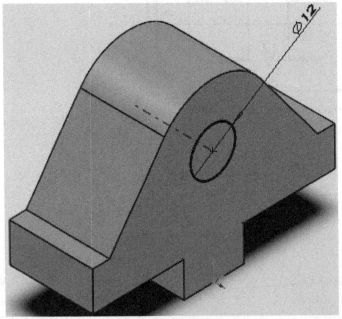

Figure 7-44 Extrude1 for Sketch1

9. Choose the right-hand side of the part.
10. Sketch a rectangle (10 mm by 36 mm) and extrude through the entire length of 64 mm.
11. Sketch a circle on the front face of diameter 12 mm and extrude-cut to a depth of 24 mm.
12. Use the Hole Wizard to insert a hole in the bottom face of diameter 6 mm, 16 mm blind.
13. Use a linear pattern to replicate the hole (2 instances and the distance is 16 mm). See Fig. 7-45.
14. Fillet the two edges where the slanting lines meet the horizontal lines.
15. Choose the top of the rectangular flat face.
16. In Sketch Mode, locate the mid-point of the 64 mm edge.
17. From the mid-point draw a construction line to the top of the arc.
18. Sketch a circle of diameter 10 mm. See Fig. 7-46.
19. Using the circle, create a countersink of diameter 10 mm, to a depth of 4 mm using an extruded cut.
20. Make the center of the 10 mm circle at the bottom of the countersink.
21. Sketch a circle of diameter 6 mm. See Fig. 7-47.
22. Create a hole from the bottom of the countersink, 6 mm deep, using the Up To Next option. See Fig. 7-48.
23. Click OK to complete the vise jaw part model.

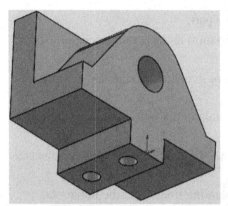

Figure 7-45 Linear-Pattern to replicate the first hole

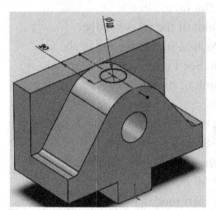 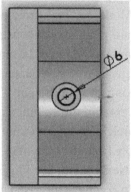

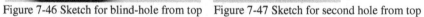

Figure 7-46 Sketch for blind-hole from top   Figure 7-47 Sketch for second hole from top

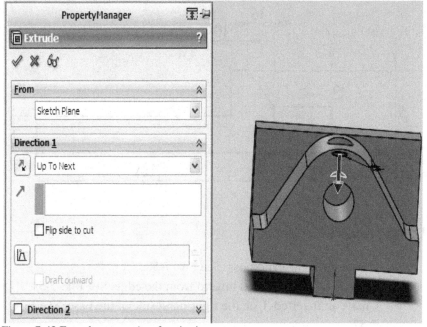

Figure 7-48 Extrude cut preview for vise jaw

Vise base

1. Choose the right-hand plane.
2. Sketch the base, Sketch1. See Fig. 7-49.
3. Extrude through 64 mm.

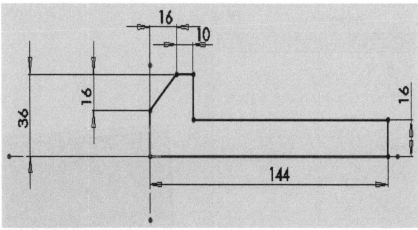

Figure 7-49 Sketch1

4. Create the right-hand boss. See Fig. 7-50.
5. Extrude the boss 22 mm inward from the edge. See Fig. 7-51.
6. Create an inverted-T on the right-hand side of the base. See Fig. 7-52.
7. Extrude-cut with All Through as the End Condition.
8. Fillet the two sides of the boss on the top of base. See Fig. 7-53.

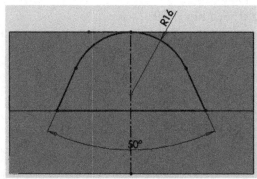

Figure 7-50 Sketch for right-hand boss

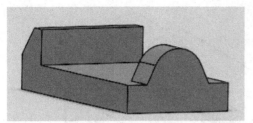

Figure 7-51 Sketch after right-hand boss has been extruded

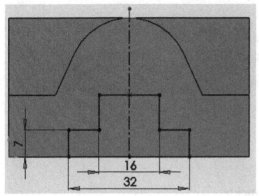

Figure 7-52 Sketch for bottom slot

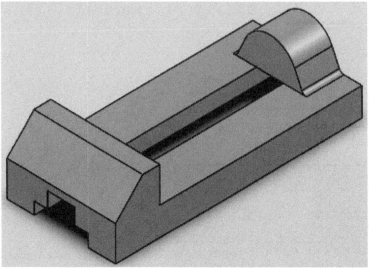

Figure 7-53 Vise base

7.7   Assembling the Components

Now that all the parts needed for assembly are ready, the first step in the assembly process is to assemble the first two components. In the bench vise assembly project, the base and vise jaw have three faces for mating. Therefore, it would easier if advantage is taken of the mate reference facility of SolidWorks. This is not always used for all assemblies; it depends on the number of faces that are mating.

Creating the mating references
1.   Click the Open button from the Menu Bar toolbar.
2.   Double-click on the vise base to open it in the SolidWorks window. This is opened in part mode.
3.   Click Feature > Reference > Reference Geometry > Mate Reference.

4. Select the top planar face, Face<1>, of the model as the Primary Reference Entity, as shown in Fig. 7-54. The selected planar face is highlighted.

5. Select the Coincident option from the Mate Reference Type drop-down list (of Tangent, Concentric, Coincident, and Parallel) in the Primary Reference Entity rollout.

6. Select the slot planar face, Face<2>, of the model as the Secondary Reference Entity, as shown in Fig. 7-54. The selected planar face is highlighted as shown.

7. Select the Coincident option from the Mate Reference Type drop-down list in the Secondary Reference Entity rollout.

8. Select the vertical planar face, Face<3>, of the model as the Tertiary Reference Entity, as shown in Fig. 7-54. The selected planar face is highlighted as shown.

9. Select the Parallel option from the Mate Reference Type drop-down list in the Tertiary Reference Entity rollout.

10. Enter Vise_Mate_Reference as the name of the mate reference in the edit box in the rollout.

11. Open the vise jaw.

12. In the same way as for the vise base, define the reference entities. The faces to be selected are shown in Fig. 7-55. Enter Vise_Mate_Reference as the name of the mate reference.

13. At this point, the only part documents that should be open are the vise base (ViceBase.sldprt) and the vise jaw (ViceJaw.sldprt). We are now ready to assemble the first two components.

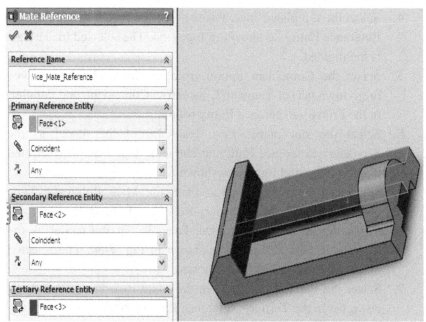

Figure 7-54 Primary, secondary, and tertiary reference entities for the vise base

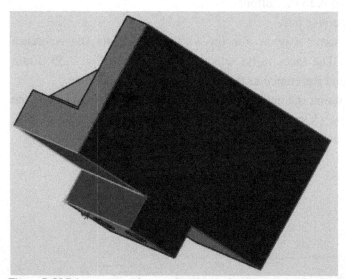

Figure 7-55 Primary, secondary, and tertiary reference entities for the vise jaw part

Assembling the first two components (base and vise jaw)

1. From either part document (vise base or vise jaw), click the New dialog. Click Make Assembly from Part/Assembly. The Begin Assembly PropertyManager pops up. The names of the components that are open are displayed in the Open Document selection box (see Fig. 7-56(a)).

2. Select the vice base as the part to be opened first.

3. Click the OK button from the Begin Assembly PropertyManager. This places the first component (vise base) of the assembly at the origin.

4. Click Assembly > Insert Components. Select the vise jaw from the Open documents selection box, as shown in the preview in Fig. 7-56(b).

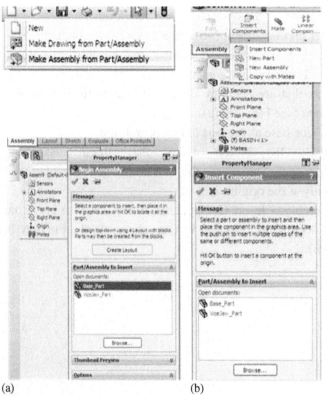

(a)   (b)

Figure 7-56 Starting an assembly

5. Place the inserted component at the required location in relation to the first component. The mates already specified are applied to the first two components. Fig. 7-57 shows the first two assembled components.

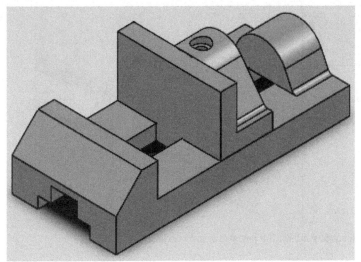

Figure 7-57 First two assembled components, vise base and vise jaw

Assembling the jaw screw

1. Click the Insert Components button from the Assemble CommandManager.

2. Click the Browse button in the Part/Assembly to Insert rollout. The Open dialog appears.

3. Double-click on the vise jaw from the Open documents selection box as shown in the preview.

4. Place the inserted component in any location in the drawing area.

5. Right-click the outer surface of the jaw screw and hold the Ctrl key down while selecting the inner face of the hole in the vise base, as shown in Fig. 7-58.

6. Apply the concentric relation, which is automatically chosen as the default. Click OK.

7. Apply a coincident mate between the planar faces of the jaw screw and the vise jaw as shown in Fig. 7-59. The result of this sub-assembly process is shown in Fig. 7-60.

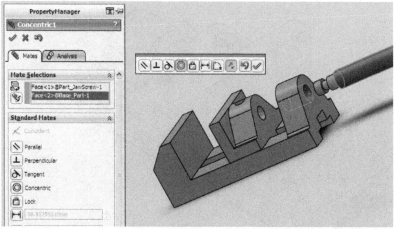

Figure 7-58 Concentric mate for the jaw screw and vise base

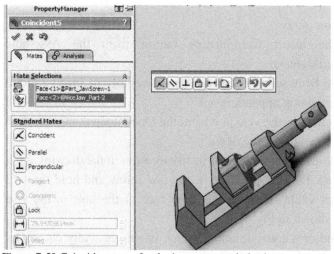

Figure 7-59 Coincident mate for the jaw screw and vise jaw

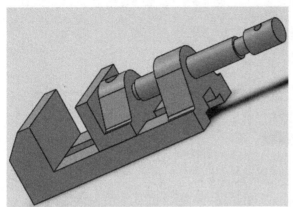

Figure 7-60 Sub-assembly of first three components

8. Select a face of the vise base and a face of the vise jaw.

9. Apply a distance mate between the two planar faces of 10 mm, as shown in Fig. 7-61.

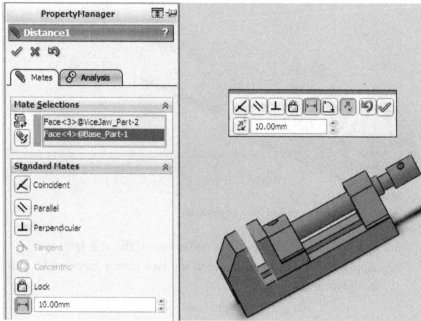

Figure 7-61 Distance mate for the vise base and vise jaw

Assembling the clamping plate

1. Click the Insert Components button from the Assemble CommandManager.
2. Click the Browse button in the Part/Assembly to Insert rollout. The Open dialog appears.
3. Double-click on the clamping plate from the Open documents selection box.
4. Rotate the sub-assembly upside down and drag the inserted component towards the holes of the vise jaw (see Fig. 7-62(a)).
5. Apply a concentric mate between the holes of the clamping plate and the vise jaw as shown in Fig. 7-62(b).
6. Apply a coincident mate between the faces of the clamping plate and the vise jaw.

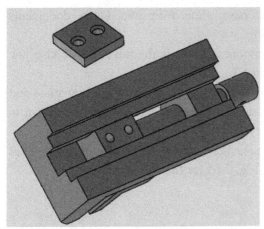

(a) Clamping plate and sub-assembly

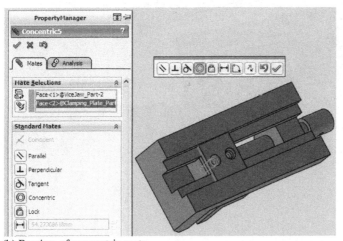

(b) Preview of concentric mate
Figure 7-62 Concentric mate between the holes of the clamping plate and vise jaw

Assembling the base plates

1. Click the Insert Components button from the Assemble CommandManager.
2. Click the Browse button in the Part/Assembly to Insert rollout. The Open dialog appears.

3. Double-click on the base plate from the Open documents selection box as shown.

4. Drag the inserted component towards the holes of the vise jaw from below. See Fig. 7-63.

5. Apply a concentric mate between the holes of the base plate and the vise jaw. Do the same for a second base plate, as shown in Fig. 7-64.

6. Apply a coincident mate between the faces of both base plates and the vise jaw. See Fig. 7-65 for the sub-assembly.

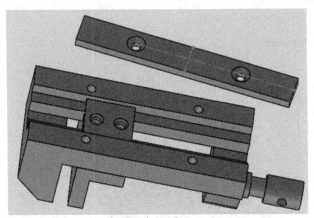

Figure 7-63 Inserting the first base plate

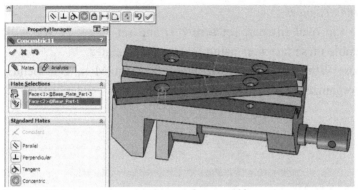

Figure 7-64 Preview of concentric mate for second base plate

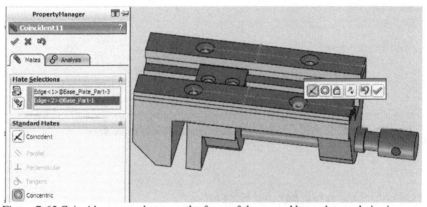

Figure 7-65 Coincident mates between the faces of the second base plate and vise jaw

Assembling the fasteners

1. Open the oval fillister, set screw 1, and set screw 2. We will assemble these fasteners using feature-based mates.
2. Choose Window > Tile Horizontally from the Menu Bar menus to reorganize the windows. See Fig. 7-66.

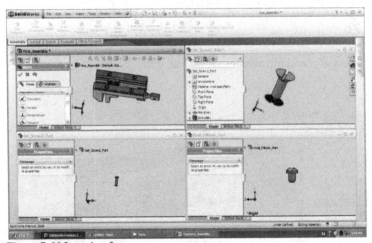

Figure 7-66 Inserting fasteners

Assembling the screw bar and bar globes

1. Click the Browse button in the Part/Assembly to Insert rollout. The Open dialog appears.
2. Double-click on the screw bar from the Open documents selection box.
3. Drop it anywhere on the graphics window.

4. Right-click the outer surface of the screw bar and hold the Ctrl key down while selecting the inner face of the hole in the jaw screw.

5. Apply a concentric relation, which is automatically chosen as the default. Click OK.

6. Carry out a similar process between the screw bar and two bar globes. The assembly is now complete as shown in Fig. 7-67.

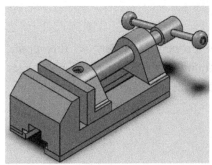

Figure 7-67 Complete assembly

7.8 Assembly Analysis

It is necessary to determine how good an assembly is, by carrying out interference analysis. This will highlight any hidden problems with the assembly, and must be carried out before any of the individual parts are machined.

7.9 Exploded View

1. Click the Exploded View Assembly Tool (see Fig. 7-68).

For each part to explode:
2. Click the part.
3. Drag the blue manipulator handle in the appropriate direction (see Fig. 7-69).
4. Click Done.

Next
5. Click OK from the Explode PropertyManager. At this point, ExplView1 is automatically found in the ConfigurationManager of the assembly. Fig. 7-70 shows the exploded view.

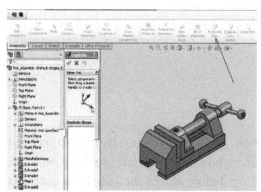

Figure 7-68 Exploded View Assembly Tool

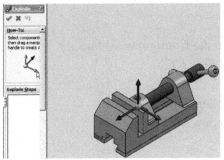

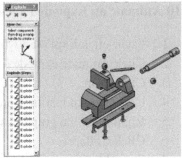

Figure 7-69 Exploding each part   Figure 7-70 Exploded view

To remove an exploded view

1. Right-click the graphics window.
2. Click Collapse (see Fig. 7-71).

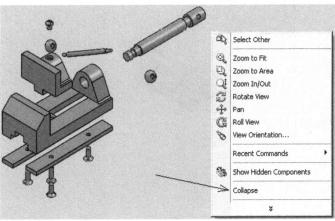

Figure 7-71 Collapsing parts in an exploded view

7.10 Animate an Exploded View

1. Right-click ExplView1 from the ConfigurationManager.
2. Click Animate explode (see Fig. 7-72).
3. Click Stop.
4. Close the Animation Controller (see Fig. 7-73).
5. Right-click the graphics window.
6. Click Collapse.
7. Save the document.

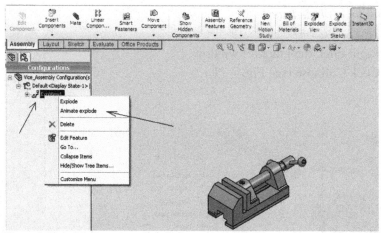

Figure 7-72 Animating an exploded view

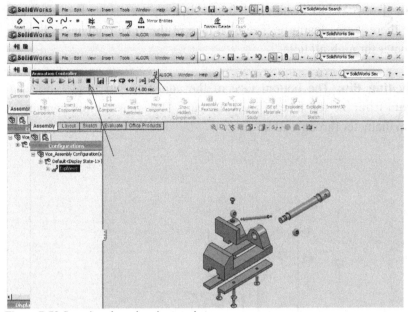

Figure 7-73 Stopping the animation session

7.11 Tutorials

Tutorial 7.1

We will model the parts and build the vise assembly shown in Fig. 7-74. The assembly contains three parts: base, yoke, and adjusting pin. Note the origin.

Units: MMGS
Material: 1060 alloy for all parts
Density: 0.0027 g mm^{-3}
Note: Fully constrain all sketches. Fully mate all parts.

Base: The distance between the front face of the base and the front face of the yoke is 60 mm.

Yoke: The yoke fits inside the left and right square channels of the base part; there is no clearance. The tope face of the yoke contains a 12 mm diameter through-all hole.

Adjusting Pin: The bottom face of the adjusting pin head is located 40 mm from the top face of the yoke part. The adjusting pin contains a 5 mm diameter through-all hole.

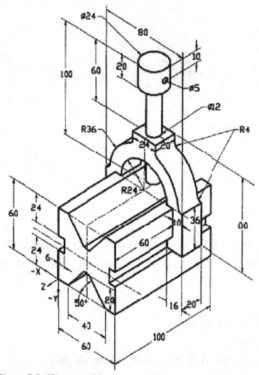

Figure 7-74 Vise assembly

Base

1. Create a new part in SolidWorks.
2. Set the document properties for the model.

3. Create Sketch1 with overall dimensions of 60 mm by 60 mm.

4. Create Extrude1 by extruding Sketch1 through a depth of 100 mm.

Yoke

5. Create a new part in SolidWorks.

6. Set the document properties for the model.

7. Create Sketch1 with overall dimensions of 80 mm by 80 mm.

8. Create Extrude1 by extruding Sketch1 through a depth of 20 mm.

Pin

9. Create a new part in SolidWorks.

10. Set the document properties for the model.

11. Create Sketch1, which is cylindrical. The shorter section has a diameter of 24 mm and is 20 mm long; the longer section has a diameter of 12 mm and is 80 mm long.

12. Create Revolved1 by revolving Sketch1 about the vertical axis.

The base, yoke, and pin are shown in Fig. 7-75.

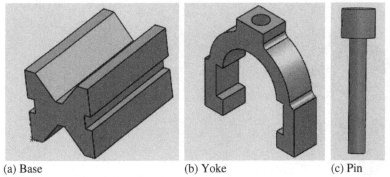

(a) Base   (b) Yoke   (c) Pin

Figure 7-75 Part models for assembly

Assembly modeling

13. Create a new assembly in SolidWorks.

14. Click Cancel from the Begin Assembly PropertyManager.

15. Set the document properties for the model.

16. Insert the base as the first part in the assembly.

17. Insert the yoke as the second part in the assembly.

18. Create parallel and concentric mates between the upper sides of the base and the inner, notched part of the yoke.

19. Create a distance mate of 60 mm between the front face of the base and the front face of the yoke.

20. Insert the pin as the third part in the assembly.

21. Create a concentric mate between the pin and the yoke.

22. Create a distance mate of 40 mm between the bottom face of the adjusting pin head and the top face of the yoke.

The base, yoke, and pin assembly is shown in Fig. 7-76.

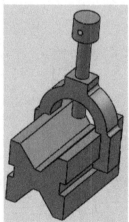

Figure 7-76 Assembly of base, yoke, and pin

Tutorial 7.2

The assembly shown in Fig. 7-77 contains three machined brackets and two pins. The assembly origin is as shown.
  Units: MMGS
  Decimal places: 2
  Note: Fully constrain all sketches.

Brackets: The brackets are identical, are 2 mm in thickness, and the holes are through all. The top edge of the notch is located 20 mm from the top edge of the machined bracket.
  Material: 6061 alloy
  Density: 0.0027 g mm^{-3}

Pins: The pins are identical and are 5 mm in length. They are mated concentric to the bracket holes with no clearance. Their end faces are coincident to the outer faces of the brackets. There is a 1 mm gap between the brackets. The brackets are positioned with equal angle mates (45°).
  Material: Titanium
  Density: 0.0046 g mm^{-3}

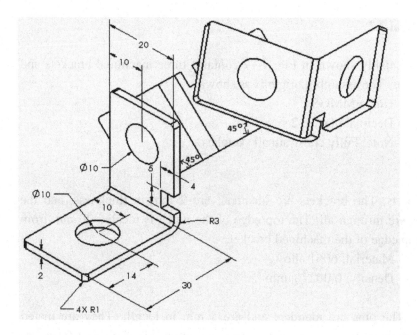

Figure 7-77 Bracket-pin assembly

Bracket

1. Create a new part in SolidWorks.

2. Select the right-hand plane as the sketch plane.

3. Create Sketch1 as an L-shape, with both arms 30 mm long.

4. Create Extrude1 by extruding Sketch1 through a depth of 20 mm.

5. Create Sketch2, the notch in the bracket, as a rectangle (4 mm by 5 mm).

6. Create Extrude2 by applying an extruded cut to Sketch2 through the bracket's thickness of 2 mm.

7. Create Sketch3, the hole in the bracket, as a circle (10 mm in diameter).

8. Create Extrude3 by applying an extruded cut to Sketch3 through the bracket's thickness of 2 mm. Repeat for the other hole.

9. Create fillets at each of the four corners with a radius of 1 mm; and create fillets of radii 3 mm and 1 mm for the outer and inner bends of the bracket, respectively. See Fig. 7-78(a).

Pin

10. Create a new part in SolidWorks.

11. Select the right-hand plane as the sketch plane.

12. Create Sketch1 as a cylinder (10 mm in diameter).

13. Create Extrude1 extruding Sketch1 through a length of 5 mm. See Fig. 7-78(b).

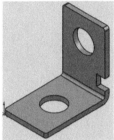

(a) Bracket (right view)   (b) Pin (front view)
Figure 7-78 Part models for the machined bracket and pin

Assembly modeling

14. Create a new assembly in SolidWorks.

15. Click Cancel from the Begin Assembly PropertyManager.

16. Set the document properties for the model.

17. Insert the bracket as the first part in the assembly.

18. Insert the pin as the second part in the assembly.

19. Create a concentric mate between the pin and the hole in the vertical part of the bracket.

20. Create a distance mate of 1 mm between the brackets.

21. Create an angular mate of 45° between the edges of the first two brackets.

22. Insert another bracket as the third part in the assembly.

23. Insert another pin as the fourth part in the assembly.

24. Create a concentric mate between the pin and the hole in the inclined part of the bracket.

25. Create a distance mate of 1 mm between the brackets.

26. Create an angular mate of 45° between the edges of the last two brackets.

The assembly of the brackets and pins is shown in Fig. 7-79. The FeatureManager for the mating of the parts is shown in Fig. 7-80.

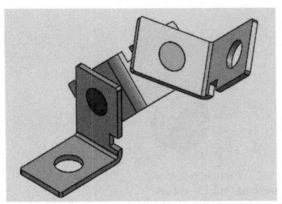

Figure 7-79 Assembly model for the machined brackets and pins

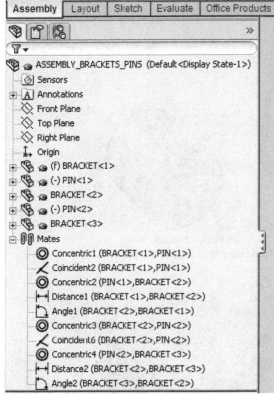

Figure 7-80 FeatureManager showing the mating of the parts in the assembly

Tutorial 7.3

Repeat Tutorial 7.2, but with straight edges for the simple brackets, a distance mate of 1 mm between the brackets, and an angular mate of 45° between the edges of the brackets.

The bracket and pin are shown in Fig. 7-81.

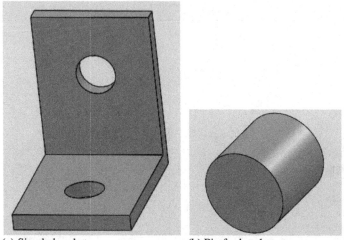

(a) Simple bracket   (b) Pin for bracket

Figure 7-81 Part models for the simple machined bracket and pin

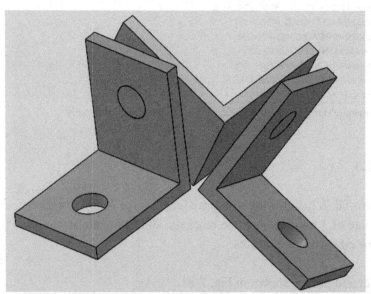

Figure 7-82 Assembly of simple bracket and pins

7.12 Project

The tailstock assembly shown in Fig. 7-83 contains five machined parts. SolidWorks files for the parts are available. Build and assemble the parts. Apply a distance mate between the front flat face of the wedge and the side flat face of the block of 0.625 in. Apply a distance mate between one side flat face of the wedge and the inner face of the hole of the post of 0.035 in.

Produce an exploded view of the assembly, a bill of materials (BOM), and a drawing for each of the five parts, including all necessary views.

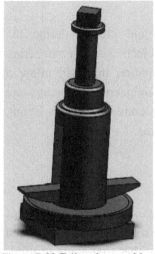

Figure 7-83 Tailstock assembly

Chapter 8

Part and Assembly Drawings

Objectives:

When you complete this chapter you will be able to:

- Understand the differences between a part drawing and an assembly drawing
- Apply the bottom-up approach to assembly modeling
- Create a new drawing template
- Insert and position views on a drawing
- Use the Exploded View Assembly Tool
- Animate the exploded view and create a bill of materials

8.1 Introduction

An engineering drawing is used to fully and clearly define the requirements for engineered items. The main objective of an engineering drawing is to convey all the required information that will allow a manufacturer to produce that component. Engineering drawings are usually created in accordance with standard conventions for layout, nomenclature, interpretation, appearance, and size.

8.2 Orthographic Projection

An orthographic projection shows the object as it looks from six sides (front, top, bottom, right, left, and back) and, in a drawing, are positioned relative to each other according to the rules of either first-angle or third-angle projection (see Fig. 8-1).

(a) First angle   (b) Third angle

Figure 8-1 Orthographic projections

First-angle projection is the ISO standard and is primarily used in Europe. The 3D part is projected onto 2D paper as if one were looking at X-ray photographs of the object so that the top view is positioned under the front view and the right-hand view is to the left of the front view (see Fig. 8-2).

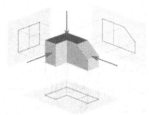

Figure 8-2 First-angle projection of a part

Third-angle projection is primarily used in the United States and Canada, where it is the default projection system according to BS 8888 (2006). The left-hand view is placed on the left and the top view on the top.

Drawing sizes

The size of a drawing normally complies with either of two different standards, the ISO (World Standard) or US customary, as shown in Table 8-1.

Table 8-1 ISO (World Standard) and US customary drawing sizes

ISO Drawing Sizes (mm)		US Customary Drawing Sizes (in)	
A4	210 x 297	A	$8.5'' \times 11''$
A3	297 x 420	B	$11'' \times 17''$
A2	420 x 594	C	$17'' \times 22''$
A1	594 x 841	D	$22'' \times 34''$
A0	842 x 1189	E	$34'' \times 44''$

8.3 Creating SolidWorks a Drawing Template

A drawing template is made up of three components: document properties, sheet properties, and a title block.

8.3.1 *Document properties*

1. Click New.
2. Double-click Drawing.
3. Click Cancel from the Model View PropertyManager. If the Start when creating new drawing is checked, the Model View PropertyManager is selected by default (see Fig. 8-3).
4. Click Options > Document Properties from the menu bar.
5. Click Drafting Standard > ANSI.
6. Click Annotations > Font > Units, enter a font height of 3 mm and click OK (see Fig. 8-4).
7. Click Dimensions, enter 1 mm, 3 mm, and 6 mm for arrow dimensions, and click OK (see Fig. 8-5).
8. Click View Labels > Section, enter 2 mm, 6 mm, 12 mm for the Section/view size, and click OK (see Fig. 8-6).
9. Click Units > MMGS, select 0.12 as the basic unit length, and None for the basic unit angle (see Fig. 8-7).
10. Click Layer Properties, and set the layers on. If not active, right-click the CommandManager and activate it (see Fig. 8-8).
11. Click Systems Options > File Locations and add the paths, then click OK three times (see Fig. 8-9).

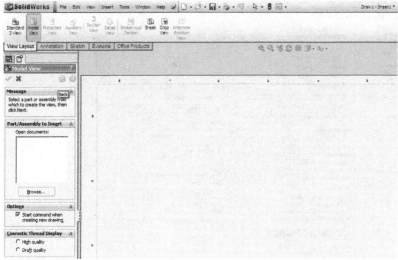

Figure 8-3 Click Cancel from the Model View PropertyManager when creating a template

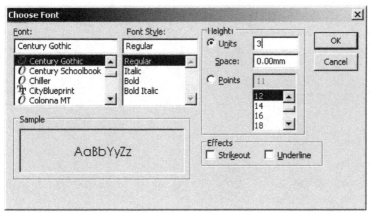

Figure 8-4 Annotations > Font > Units

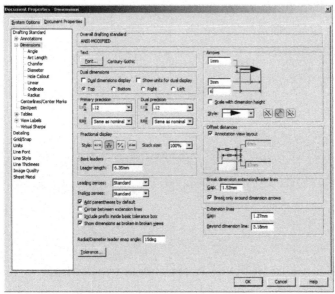

Figure 8-5 Dimensions

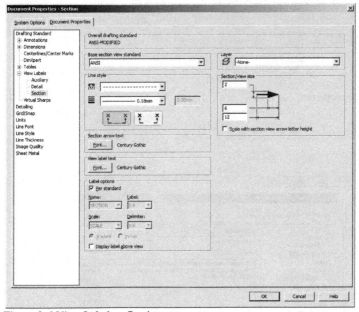

Figure 8-6 View Labels > Section

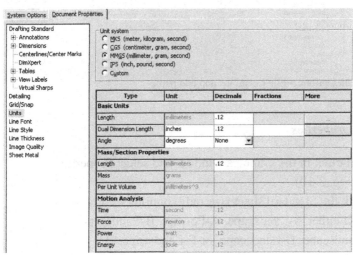

Figure 8-7 Units > MMGS

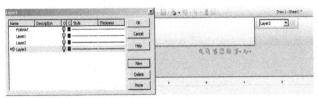

Figure 8-8 Layer properties

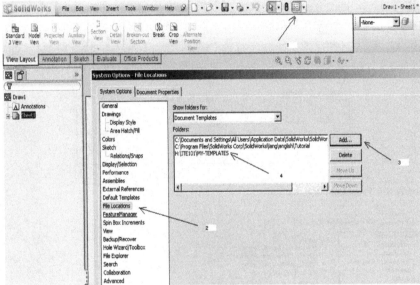

Figure 8-9 System Options > File Locations and add paths

8.3.2 *Sheet properties*

1. Right-click in the graphics window and select Properties (see Fig. 8-10).
2. Check the Standard sheet size radio button (see Fig. 8-11).
3. Select either A-Landscape, A-Portrait, B-Landscape, C-Landscape, D-Landscape, or E-Landscape.
4. Check Third Angle and set the scale as 1:1.
5. Uncheck Display sheet format.
6. Click OK.

358

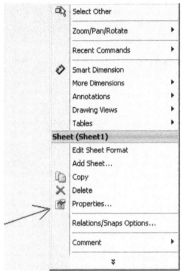

Figure 8-10 Right-click in the graphics window and select Properties

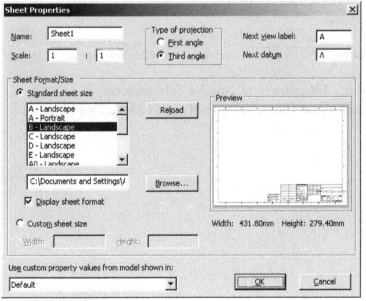

Figure 8-11 Sheet Properties

8.3.3 *Title block*

1. Right-click in the graphics window.
2. Select Edit Sheet Format (see Fig. 8-12).

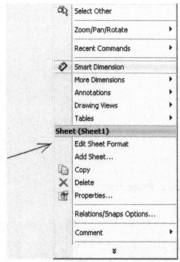

Figure 8-12 Right-click in the graphics window and select Edit Sheet Format

3. Modify Tolerance Note, Angular Tolerance, etc.
4. Create a new Microsoft logo and paste into the Title Block.

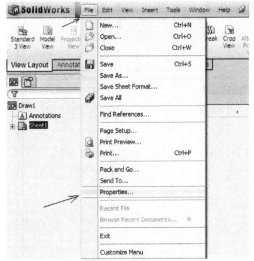

Figure 8-13 File > Properties

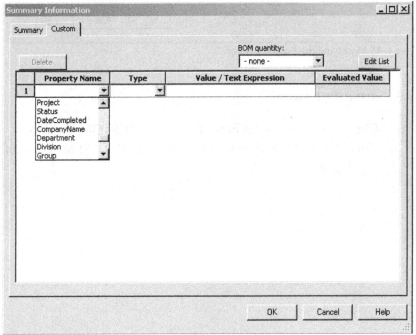

Figure 8-14 Properties > Custom and select Property Name

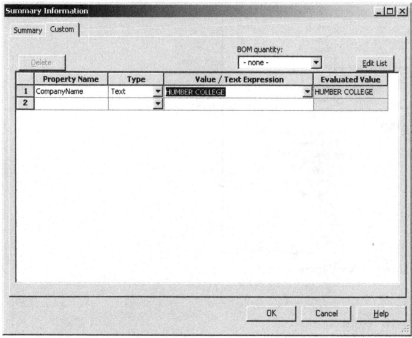

Figure 8-15 Modifying the Property Name

8.3.4 *Saving the template*

1. Click File > Save Sheet Format and enter CUSTOM-B.slddrt.
2. Click Save As and enter the filename B-ANSI-MM.drwdot (see Fig. 8-16).

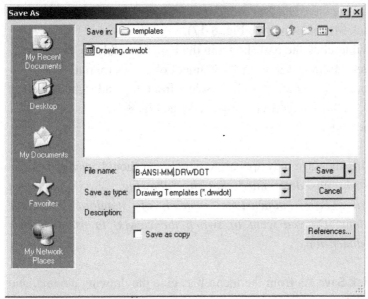

Figure 8-16 Saving a drawing template

8.4 Part Drawing of Vise Base

This section describes the steps required to generate a part drawing, using an example. Standard views are first obtained followed by the exploded view and bill of material (BOM).

We will produce a drawing of the base of the vise that was modeled in Tutorial 7.1.

8.4.1 *Standard views*

Display the Drawing FeatureManager
1. Click New from the menu bar.
2. Double-click B-ANSI-MM from the MY-TEMPLATE tab.
3. Click Cancel from the Model View PropertyManager, if it appears.

Insert a drawing view and four standard views

4. Click Model View (see Fig. 8-17).
5. Double-click the part shown in the Part/Assembly to Insert box.
6. Click Multiple Views in the Number of Views rollout.
7. From the Orientation rollout, select front, top and right as well as 3D, as the standard views needed (see Fig. 8-18).
8. Click OK.

The following message may appear, to which you should answer Yes:
SolidWorks has determined that the following view(s) may need Isometric (True) dimensions instead of standard Projected dimensions. Do you want to switch the view(s) to use Isometric (True) dimensions?

9. Click Save As from the menu bar, give the drawing a name, and click Save.

Figure 8-17 Model View for drawing session

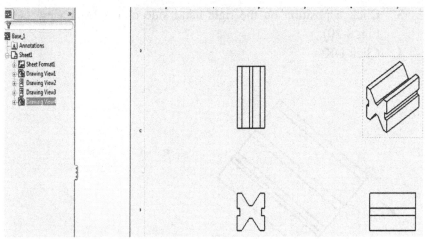

Figure 8-18 Multiple views selected

8.4.2 *Import dimensions*

1. Click Annotation > Model items.
2. Select Entire model for Source/Destination (see Fig. 8-19).
3. Click OK.

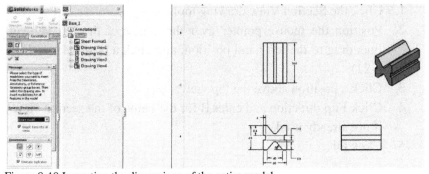

Figure 8-19 Importing the dimensions of the entire model

8.4.3 *Auxiliary View*

1. Click View Layout and choose the Auxiliary View drawing tool.
2. Click the right-angled edge of the part in the front view.

3. Click a position on the right-hand side of the front view (see Fig. 8-20).
4. Click OK.

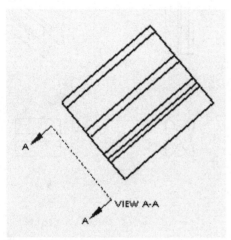

Figure 8-20 Auxiliary View

8.4.4 *Sectional View*

1. Click the Section View drawing tool.
2. Position the mouse pointer over the center of the left vertical line, drag to the left-hand position, and click a position (see Fig. 8-21).
3. Click a position above the top view.
4. Click Flip direction and enter B for the name of the section view, if not already used.
5. Click OK.

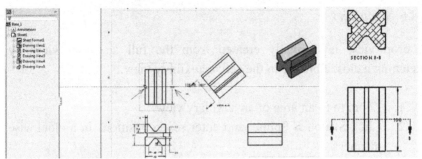

Figure 8-21 Sectional View

8.4.5 *Detail View*

1. Click the Detail View drawing tool.
2. Click a point for the inset circle (see Fig. 8-22).
3. Click a position to the right of the section view, and enter C for the name of the view.
4. Change the scale using Custom Scale > User Defined, and enter 3:2.
5. Click OK.

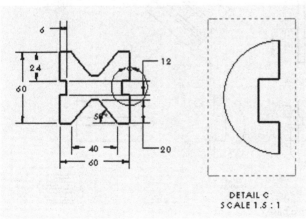

Figure 8-22 Detail View

8.4.6 *Crop View*

A crop view is normally created from the full auxiliary view by sketching a closed profile in the active auxiliary view.

1. Zoom in to an area of an auxiliary view.
2. Click Sketch > Spline, and enter seven positions in a clockwise manner.
3. Click Crop View.
4. Click OK.

See Fig. 8-23 for the final drawing of the vise base.

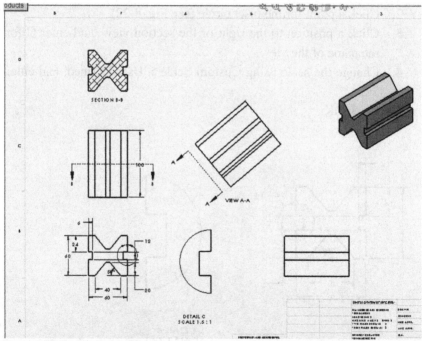

Figure 8-23 Drawing of vise base

8.4.7 *Detailed drawing*

Detailed drawing requires the following activities:
- More dimensions
- Add dimensions to holes and hole callout
- Add marks and centerlines
- Additional features, etc.
- Add notes

8.5   Assembly Drawing of Vise

As an example of an assembly drawing, we will produce the exploded view and bill of materials (BOM) for the vise assembly, modeled in Tutorial 7.1.

8.5.1 *Exploded view*

Two views must be open.

Open the assembly document already completed
1. Click Windows > Close All.
2. Open the assembly (see Fig. 8-24).

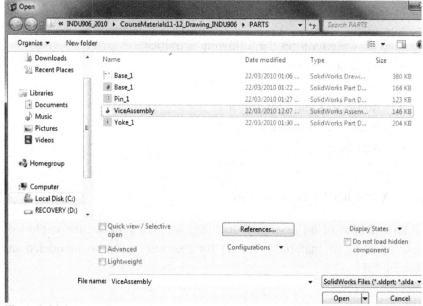

Figure 8-24 Opening the assembly

Create a new drawing

3. Click Make Drawing from Part/Assembly from New (see Fig. 8-25).

4. Double-click the B-ANSI-MM template from the MY-TEMPLATES tab.

5. Click the View Palette tool from the task pane (see Fig. 8-26).

6. Click and drag an isometric view from the View Palette to Sheet1 (see Fig. 8-27).

7. Click OK.

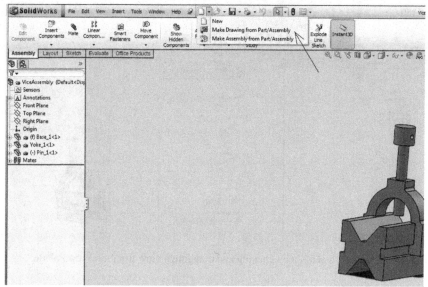

Figure 8-25 Making a drawing from an assembly

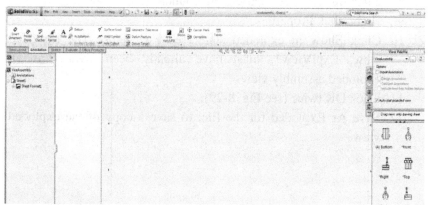

Figure 8-26 View Palette

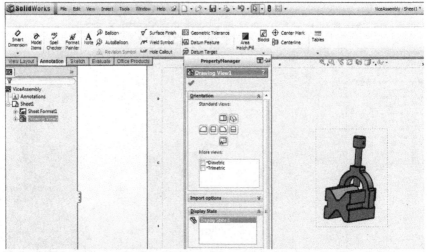

Figure 8-27 Import assembly by clicking and dragging a view from the View Palette

Display the Exploded View

8. Click inside the isometric view boundary.

9. Right-click Properties.

10. Click Show in exploded state (see Fig. 8-28). The exploded view, ExplView1, must have already been created as an exploded assembly view.

11. Click OK twice (see Fig. 8-29).

12. Save As Exploded for the file, to save a copy of the exploded view.

Figure 8-28 Checking Show in exploded state

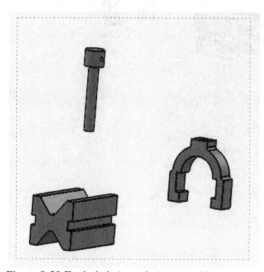

Figure 8-29 Exploded view of vise assembly

8.5.2 *Balloons*

1. Click inside the isometric view boundary (see Fig. 8-30).
2. Click Auto Balloon from the Annotation toolbar (see Fig. 8-31).
3. Click OK.
4. Select Balloon Settings > Select Circular Split Line (see Fig. 8-32).
5. Click OK.
6. Save.

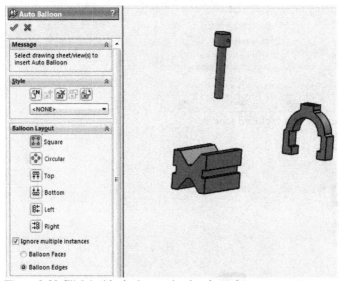

Figure 8-30 Click inside the isometric view boundary

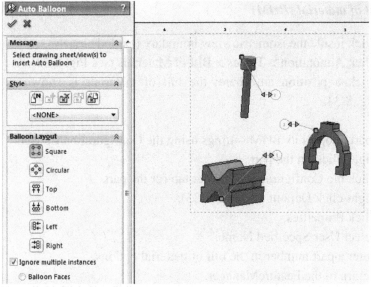

Figure 8-31 Click Auto Balloon

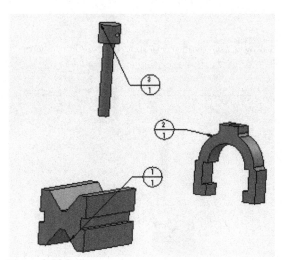

Figure 8-32 Balloons

8.5.3 *Bill of material (BOM)*

1. Click inside the isometric view boundary (exploded view).
2. Click Annotation > Tables > Bill of Materials (see Fig. 8-33).
3. Click a position anywhere; the bill of materials is shown in Fig. 8-34.

For each part, modify its BOM settings using the ConfigurationManager:

1. Right-click on the part.
2. Click the ConfigurationManager tab for the part.
3. Right-click Default (see Fig. 8-35).
4. Click Properties.
5. Select User Specified Name.
6. Enter a part number in the bill of material options.
7. Return to the FeatureManager.

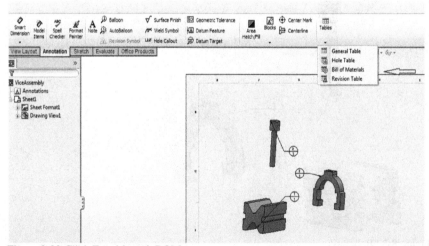

Figure 8-33 Click Excel-based, BOM

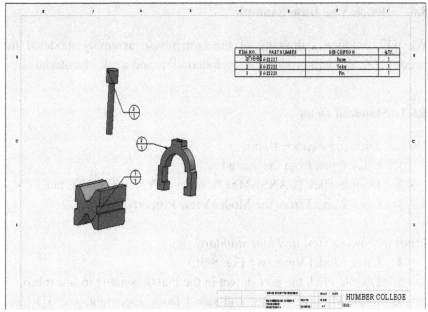

ITEM NO.	PART NUMBER	DESCRIPTION	QTY.
1	56-22221	Base	1
2	56-22222	Yoke	1
3	56-22220	Pin	1

HUMBER COLLEGE

Figure 8-34 BOM

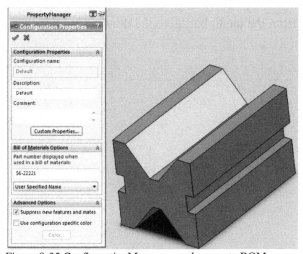

Figure 8-35 ConfigurationManager used to create BOM

377

8.6   Bench Vise Base Example

We will produce a drawing of the bench vise assembly modeled in Section 7.6. We will produce an exploded view and a bill of materials.

8.6.1 *Standard views*

Display Drawing FeatureManager
   1.   Click Open from the menu bar.
   2.   Double-click B-ANSI-MM from the MY-TEMPLATE tab.
   3.   Click Cancel from the Model View PropertyManager.

Insert a drawing view and four standard views
   4.   Click Model View (see Fig. 8-36).
   5.   Double-click the part shown in the Part/Assembly to Insert box.
   6.   Click Multiple Views, and select front, top, right, and 3D (see Fig. 8-37).
   7.   Click OK.
   8.   Click Save As from the menu bar, give the drawing a name, and Save.

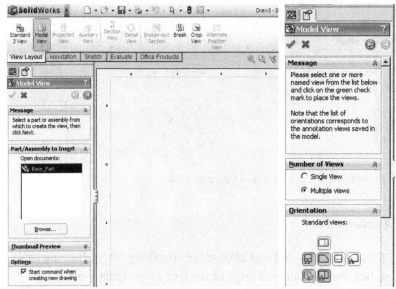

Figure 8-36 Model view for drawing session

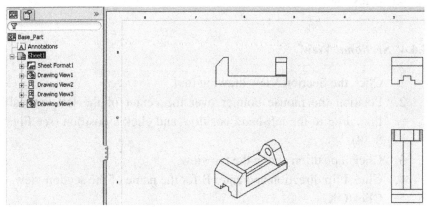

Figure 8-37 Multiple views selected

8.6.2 *Import dimensions*

1. Click Annotation > Model items.
2. Select Entire model for Source/Destination (see Fig. 8-38).
3. Click OK.

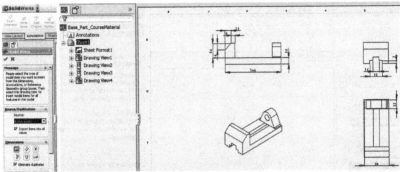

Figure 8-38 Importing the dimensions of the entire model

8.6.3 *Auxiliary View*

1. Click View Layout and choose the Auxiliary View drawing tool.
2. Click the right-angled edge of the part in the front view.
3. Click a position on the right-hand side of the front view.
4. Click OK.

8.6.4 *Sectional View*

1. Click the Section View drawing tool.
2. Position the mouse pointer over the center of the left vertical line, drag to the left-hand position, and click a position (see Fig. 8-39).
3. Click a position above the top view.
4. Click Flip direction and enter B for the name of the section view.
5. Click OK.

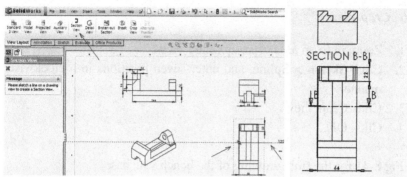

Figure 8-39 Sectional view

8.6.5 *Detail View*

1. Click the Detail View drawing tool.
2. Click a point for the inset circle (see Fig. 8-40).
3. Click a position to the right of the section view, and enter C for the name of the view.
4. Change the scale using Custom Scale > User Defined, and enter 3:2.
5. Click OK.

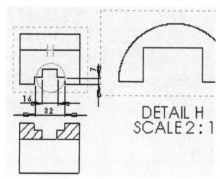

Figure 8-40 Detail View

8.6.6 *Crop View*

1. Zoom in to an area of an auxiliary view.
2. Click Sketch > Spline, and enter seven positions in a clockwise manner.
3. Click Crop View.
4. Click OK.

See Fig 8-41 for the final drawing of the bench vise base.

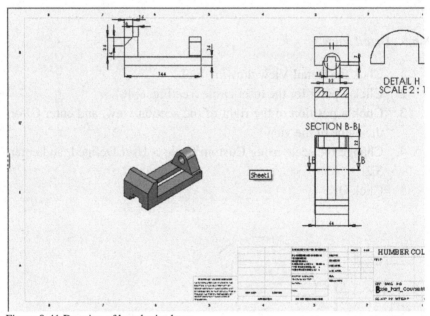

Figure 8-41 Drawing of bench vise base

8.7 Assembly Drawing of Bench Vise

8.7.1 *Exploded View*

Two views must be open.

Open the assembly document already completed
1. Click Windows > Close All.
2. Open the assembly (see Fig. 8-42).

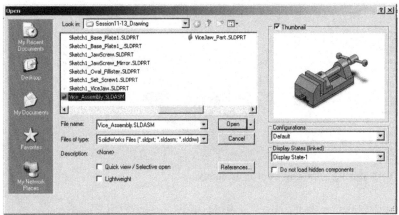

Figure 8-42 Opening the assembly

Create a new drawing
3. Click Make Drawing from Part/Assembly from New (see Fig. 8-43).
4. Double-click the B-ANSI-MM template from the MY-TEMPLATES tab.
5. Click the View Palette tool from the task pane (see Fig. 8-44).
6. Click and drag an isometric view from the View Palette to Sheet1 (see Fig. 8-45).
7. Click OK.

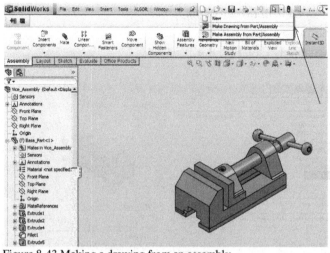

Figure 8-43 Making a drawing from an assembly

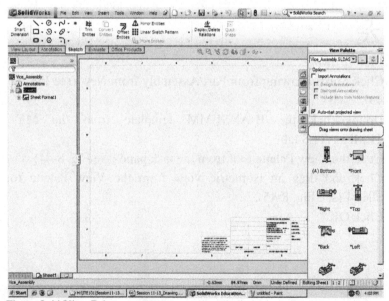

Figure 8-44 View Palette

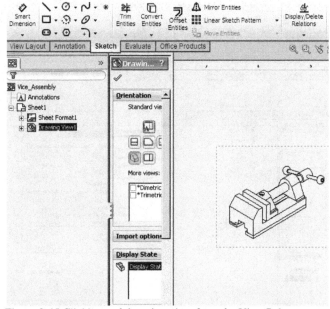

Figure 8-45 Clicking and dragging view from the View Palette

Display the Exploded View

8. Click inside the isometric view boundary.

9. Right-click Properties.

10. Click Show in exploded state (see Fig. 8-46).

11. Click OK twice (see Fig. 8-47).

12. Save As Exploded for the file.

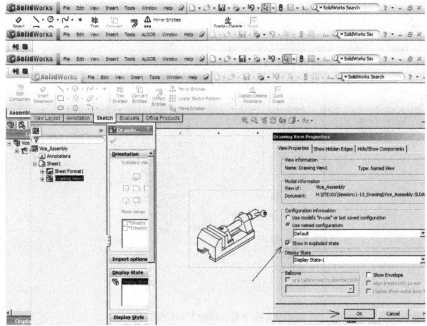

Figure 8-46 Checking Show in exploded state

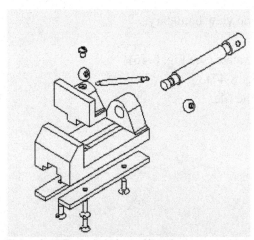

Figure 8-47 Exploded view of bench vise assembly

8.7.2 *Balloons*

1. Click inside the isometric view boundary (see Fig. 8-48).
2. Click Auto Balloon from the Annotation toolbar (see Fig. 8-49).
3. Click OK.
4. Select Balloon Settings > Select Circular Split Line (see Fig. 8-50).
5. Click OK.
6. Save.

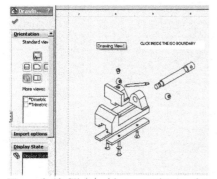

Figure 8-48 Click inside isometric view boundary

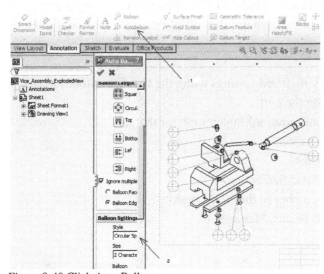

Figure 8-49 Click Auto Balloon

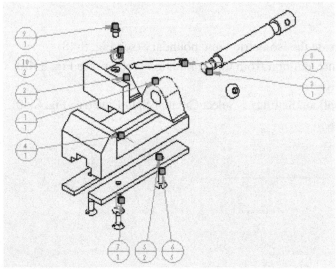

Figure 8-50 Balloons

8.7.3 *BOM*

1. Click inside the isometric view boundary (exploded view).
2. Click Annotation > Tables > Bill of Materials (see Fig. 8-51).
3. Click a position anywhere; the bill of materials is shown in Fig. 8-52.

For each part, modify its BOM settings using the ConfigurationManager:
1. Right-click on the part.
2. Click the ConfigurationManager tab for the part.
3. Right-click Default.
4. Click Properties.
5. Select User Specified Name.
6. Enter a part number in the BOM options.
7. Return to the FeatureManager.

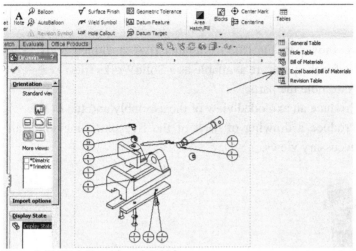

Figure 8-51 Click Excel-based, BOM

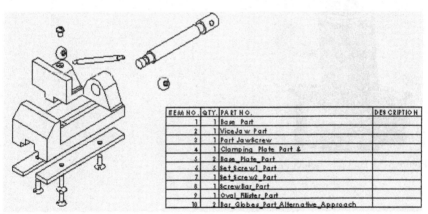

ITEM NO.	QTY.	PART NO.	DESCRIPTION
1	1	Base Part	
2	1	ViceJaw Part	
3	1	Part JawScrew	
4	1	Clamping Plate Part &	
5	2	Base_Plate_Part	
6	5	Set Screw1_Part	
7	1	Set Screw2_Part	
8	1	ScrewBar Part	
9	1	Oval_Filister_Part	
10	2	Bar_Globes_Part Alternative_Approach	

Figure 8-52 BOM

Exercise

The assembly shown in Fig. 8-53 contains five machined parts. The electronic files for the parts are available as a SolidWorks files.

1. Assemble the parts.
2. Produce an exploded view of the assembly and the BOM.
3. Produce a drawing of each of the five parts, including all necessary views.

Figure 8-53 Tool post holder

Chapter 9

Configurations and Design Tables

Objectives:

When you complete this chapter you will have:

- Learnt how create a design table that generates different configurations
- Learnt how design tables lead to more flexible designs

9.1 Design Tables and Configurations

Design tables are embedded Excel spreadsheets that drive dimension parameters, feature and part suppression, and visibility. They are powerful and flexible, making designs more flexible and modular. A designer can use configurations to create multiple versions of the same design with different features or dimension values. Alternatively, a designer can simplify a complex assembly by breaking its sub-assemblies into separate configurations.

There is no limit to their application. Whatever the task, a design table can be used.

Design tables make use of shared values. The concept is based on the fact that every dimension in a document (either a part or an assembly) is unique. We can verify this by highlighting each dimension and viewing its unique name. We will illustrate the use of shared values by means of an example.

9.2 Example: Socket Head Screw

A sketch and model of a socket head screw are shown in Fig. 9-1. The screw diameter is 30 mm, its length is 90 mm, the diameter of the screw head is 48 mm, and the height of the screw head is 30 mm. The dimensions for the across-the-parallel-faces and depth for the hexagonal head sink are 24 mm and 18 mm, respectively.

Our goal is to create a design table for different configurations of the screw, with diameters 20 mm, 50 mm, 40 mm, and 5 mm, and similarly for its other dimensions.

391

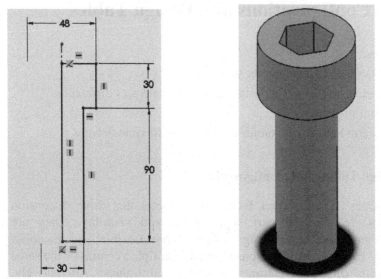

Figure 9-1 Sketch and model for which configurations are needed

9.2.1 *Part modeling*

The Revolved Boss/Base tool rotates a contour about an axis. It is a useful tool when modeling parts that have circular symmetry. We will use this tool to create the screw.

1. Select the front plane and create Sketch1, as shown in Fig. 9-2.
2. Click the Features tool.
3. Click the Revolved Boss/Base tool. The Revolve Property-Manager appears (see Fig. 9-3).
4. Define the revolved axis (Line1) as the vertical dimension line. A real-time preview will appear.
5. Click OK to complete the revolved part, as shown in Fig. 9-4.

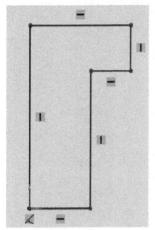 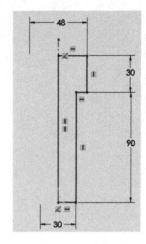

Figure 9-2 Sketch1

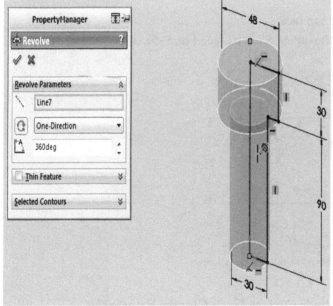

Figure 9-3 Revolve PropertyManager

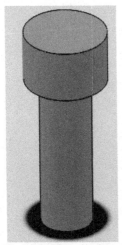

Figure 9-4 Revolved part

6. Click the top face.
7. Sketch a hexagon, Sketch2 (see Fig. 9-5).

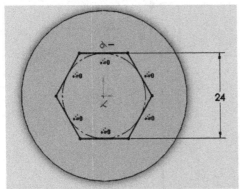

Figure 9-5 Sketch2, hexagon on top face of screw

8. Click Extrude Cut (see Fig. 9-6 for the Extrude Cut PropertyManager).
9. Select Blind in the Direction 1 rollout and set the depth to 18 mm.
10. Click OK to complete the extruded cut.

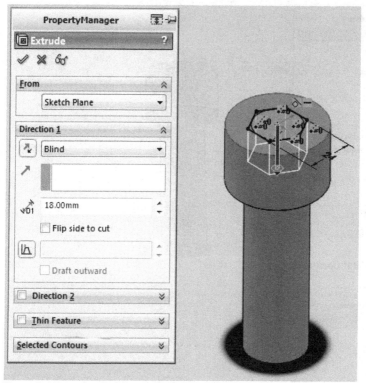

Figure 9-6 Extrude Cut PropertyManager

Filleting and chamfering

11. Click at the junction of the two cylinders (see Fig. 9-7).
12. Click Fillet and set the fillet radius to 1.20 mm with the Tangent propagation option.
13. Click OK.
14. From the Chamfer Parameters rollout, select the Angle distance option (see Fig. 9-8).
15. Set the chamfer distance to 1.20 mm and the chamfer angle as 45°.
16. Click OK.

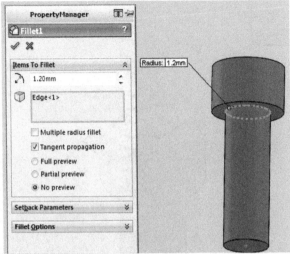

Figure 9-7 Fillet PropertyManager

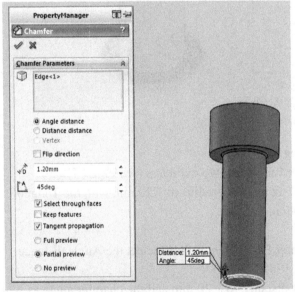

Figure 9-8 Chamfer PropertyManager

9.2.2 *Shared values*

Shared values use a variable to link model dimension values, as shown in Fig. 9-2. Shared values are based on the premise that every dimension in a document is unique. You can verify this by checking the labels or names of the dimensions in a part or assembly. You can edit any of the linked dimensions, and all linked values will be updated. You can also create more than one variable within a document. The variable can be used only within the current document.

To create shared values:
1. Click on each dimension (see Fig. 9-9).
2. Right-click on each dimension, a contextual tool appears.
3. Click Link Values.

To create a linked value, select a dimension and right-click Link Values. To create a new shared value, enter the variable name. To use an existing value, select it from the pull-down list. A red mark appears before a dimension with a linked value. To unlink a dimension, edit the sketch; select the dimension, and right-click Unlink Value.

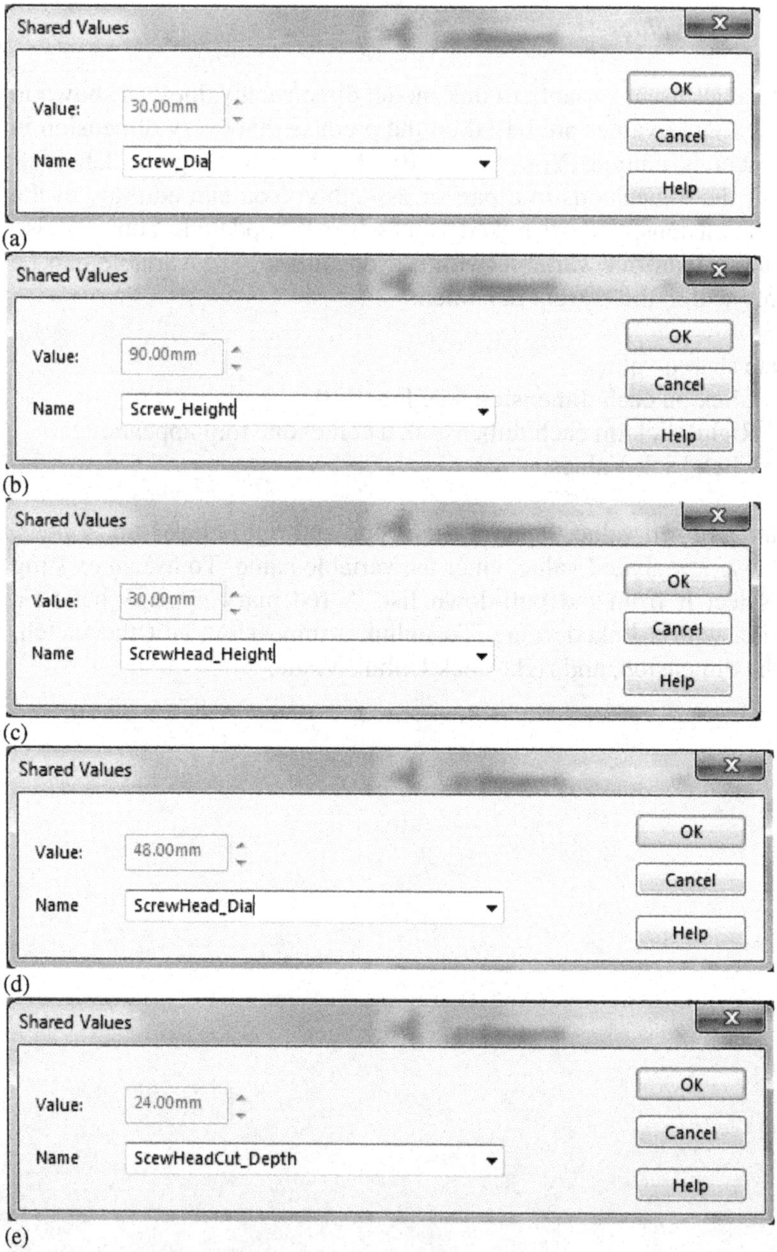

Figure 9-9 Dialogs for shared values

A designer can create configurations either manually or by using design tables. The advantage of using design tables is the ease of viewing, editing, and understanding the differences among the configurations because they can be displayed in a spreadsheet.

9.2.3 *Design Tables*

To create a design table:
1. Click Insert > Tables > Design Table.
2. Click OK. A spreadsheet appears, as shown in Fig. 9-10; a list of dimension names is also displayed.
3. To add a dimension to the table, highlight its name by holding down the Ctrl key and right-click.
4. Click OK.

Avoid highlighting the following:
*   D1@Revolve1
*   D1@Fillet1
*   D1@Chamfer1

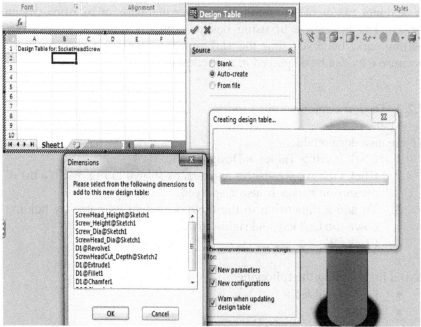

Figure 9-10 Design table created using shared values and names

A design table creates a default configuration with the original dimensions (see Fig. 9-4). We need to add a number of configurations to this design table, as shown in Fig. 9-11.

The rows in the design table show configurations for several different sizes of the socket head screw. The columns show the names of the driving dimensions. If the default names were used, it would be difficult to figure out what's being controlled.

	A	B	C	D	E	F	G
1	Design Table for: SocketHeadScrew						
2		ScrewHead_Height@Sketch1	Screw_Height@Sketch1	Screw_Dia@Sketch1	ScrewHead_Dia@Sketch1	ScewHeadCut_Depth@Sketch2	D1@Extrude1
3	Default	30	90	30	48	24	18
4	20x60	20	60	20	32	16	12
5	40x120	40	120	40	64	32	24
6	50x150	50	150	50	80	40	30
7	5x15	5	15	5	8	4	3
8							
9							
10							

Sheet1

Figure 9-11 Design table with five configurations

Design tables create part or assembly configurations. To view the defined configurations, select the ConfigurationManager tab at the top of the FeatureManager design tree. To go back to the FeatureManager mode, select the FeatureManager tab. These tabs are used to switch the mode (feature- or configuration-based) of the navigation tool. See Fig. 9-12 for the configurations in the ConfigurationManager.

Figure 9-12 ConfigurationManager shows all defined configurations

The configurations are now complete. We can use a configuration by double-clicking on it in the ConfigurationManager.

9.3 Comments

Another advantage of design tables is that you can add comments to the spreadsheet. When there is a comment embedded in a cell in the spreadsheet, a red triangle is shown in the upper right-hand corner of the cell. Move the mouse over the cell and the comment will be displayed.

In addition, you can add comments to a blank row in column A or add them to any row 1 cell.

Chapter 10

Surface Modeling

Objectives:

When you complete this chapter you will have:

- Learnt how to create freeform surfaces
- Learnt how to use control polygons for freeform surface design
- Learnt how to use control freeform surfaces using control polygons
- Learnt how to create extruded surfaces
- Learnt how to create boundary surfaces
- Learnt how to create lofted surfaces
- Learnt how to create revolved surfaces

10.1 Generalized Methodology for Freeform Surface Design

In this chapter, the generalized methodology for freeform surface design is presented. In SolidWorks, there are at least two ways of creating a freeform design: (i) extracting an existing surface from a 3D part and creating a freeform surface; (ii) creating a freeform surface from scratch. The second method is not much used because it requires a good knowledge of standard surface modeling, such as the work of Coons, Bézier surfaces, and NURBS (Non-Uniform Rational B-Spline); the treatment of which are outside the scope of this course. SolidWorks uses NURBS to create freeform surfaces.

In this section, control polygons are used for freeform surface design. A control polygon is the key to a flexible and robust freeform surface design methodology.

10.2 Control Polygon

1. Click Insert > 3D Sketch.

2. Click the front plane.
3. Sketch 3DSketch1, with a set of four (or more) control points of interest, as shown in Fig. 10-1. This is the first control polygon.
4. Create four (or more) planes: Plane1, Plane2, Plane3, and Plane4 (see Fig. 10-2(a)).
5. Exit the 3D Sketch.
6. Select Plane1, right-click and click 3D Sketch On Plane.
7. Choose Normal To, to make the plane normal.
8. On this plane, make an exact copy of the first sketch, 3DSketch1, and at the same position on Plane1 as 3DSketch1 is on the front plane. This is the second control polygon.
9. Likewise, make another three exact copies of the first sketch on Plane2, Plane3, and Plane4 (see Fig. 10-2(b)).

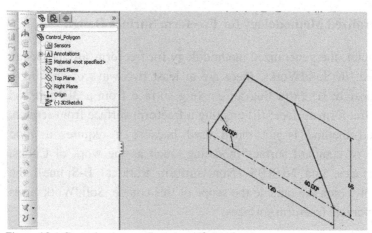

Figure 10-1 Control points on first plane (front plane)

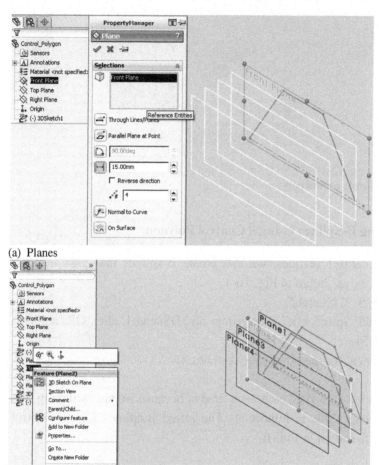

(a) Planes

(b) Sketches on Plane1 and Plane2
Figure 10-2 Planes defining control polygons

At this point, our control vertices are ready, as shown in Fig. 10-3. There are 20 vertices: four on each plane and there are five planes. Notice that there are five 3DSketches. This means that we can change each control polygon independently. This is important in surface design. For example, 3DSketch1 can be changed independently of 3DSketch2, etc.

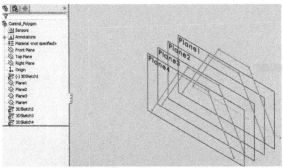

Figure 10-3 Simple control polygons

10.3 Lofting B-Splines Using a Control Polygon

Using the control vertices, we can create B-splines that pass through these vertices, as shown in Fig. 10-4.

1. Click 3DSketch1.
2. Click Spline, click the vertices on 3DSketch1, click OK, and exit 3DSketch.
3. Repeat for the other four sketches.

The reason we exit 3DSketch each time is because we are switching to a different plane. This is important! The lofted B-splines for the control polygons are shown in Fig. 10-5.

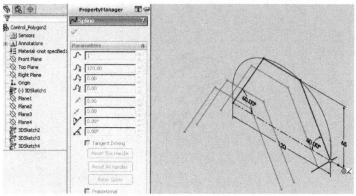

Figure 10-4 First B-spline based on control points on the first plane

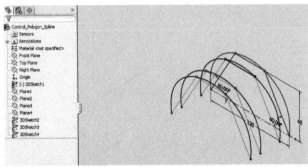

(a) Dimensions shown

(b) B-splines
Figure 10-5 B-Spline using control polygons

10.4 Freeform Surface Design

So far we have created lofted B-splines based on the control polygons. Now, let us design surfaces using the splines. We will use the boundary surface option; if we design from scratch, this is the only option that is active.

1. Click Insert > Surface > Boundary Surface.

2. Click 3DSketch_Spline1 in the Direction 1 rollout (see Fig. 10-6).
3. Click 3DSketch_Spline2.
4. Click 3DSketch_Spline3.
5. Click 3DSketch_Spline4.
6. Click 3DSketch_Spline5 (see Fig. 10-6 for a preview of the surface).
7. Click OK (see Fig. 10-7).

We have now created our first freeform surface from scratch! Notice that a new folder, Surface Bodies(1), has been created, and that FeatureManager now has a surface definition, Boundary-Surface1, which contains the splines, 3DSketch_Spline1,..., 3DSketch_Spline5.

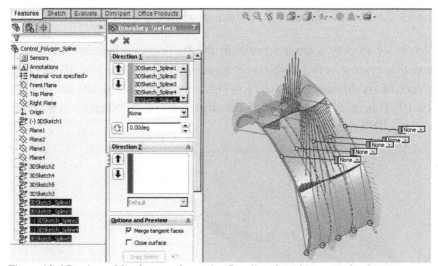

Figure 10-6 Preview of freeform surface using B-splines based on control polygons

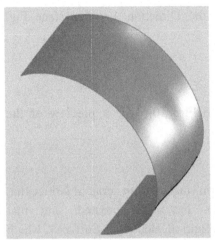

Figure 10-7 Freeform surface using B-splines based on control polygons

10.5 Effect on the Freeform Surface of Modifying a Control Polygon

When we modify a control polygon the freeform surface also changes, and this gives a very flexible way to design shapes (see Fig. 10-8). The effect of modifying the control polygon in Fig. 10-9 is shown in Fig. 10-10. This modification is to the control polygon, 3DSketch1, not to the spline, 3DSketch_Spline1.

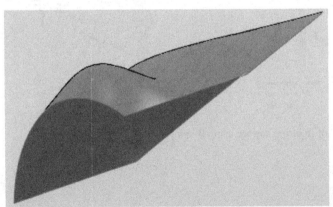

Figure 10-8 Modifying the freeform surface through modifying a control polygon

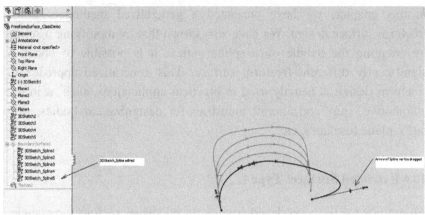

Figure 10-9 Modified control vertices

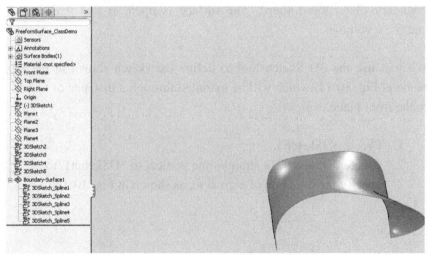

Figure 10-10 Modified shape resulting from modified control vertices

Notice that when the surface model is complete, Surface Bodies(1) is at the top in the FeatureManager, while Boundary-Surface(1) is at the bottom. The number in parenthesis is the number of surface bodies.

In this chapter, we have presented a generalized methodology for freeform surface design. We have also shown that by modifying a spline by dragging the handle on a spline vertex, it is possible to obtain a significantly different freeform surface. This generalized approach to freeform design is heavily used in practical applications, such as in the automotive, ship, and aircraft industries for designing car bodies, ship hulls, plane fuselages, etc.

10.6 Extruded Surface: Type I

An extruded solid is produced from a closed shape, such as a rectangle. An extruded surface is produced from an open shape or from the boundary of a closed shape. The surface is open so that it does not enclose a volume.

We will use the 3D Sketch tool to define the sketch shown in the top plane in Fig. 10-11, which will be extruded through a distance of 65 mm in the front plane.

1. Create 3DSketch1.
2. Create 3DSketch2, a straight line vertical to 3DSketch1, to specify the direction of extrusion, as shown in Fig. 10-12.

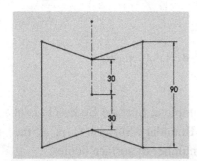

Figure 10-11 Sketch1 for creating an extruded surface

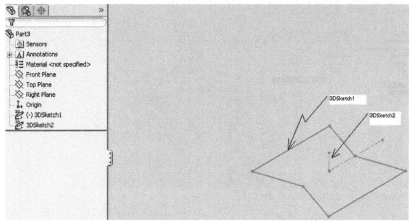

Figure 10-12 Vertical line: direction for extrusion

3. Choose 3DSketch1 in the FeatureManager.

4. Click Insert > Surface > Extrude.

5. From the Selected Contours rollout, highlight 3DSketch1 in the graphics window.

6. From the Direction 1 rollout, click 3DSketch2. A preview pops up (see Fig. 10-13(a)).

7. Enter the value of the extrusion in Direction 1 as 65 mm.

Fig. 10-13(b) shows the extruded surface.

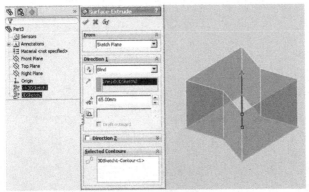

(a) Preview of the extruded surface

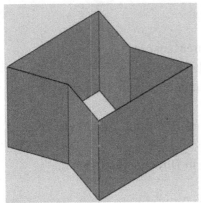

(b) Extruded surface

Figure 10-13 Extruded surface

10.7 Revolved Surface: Type I

A revolved surface uses just the boundary of a shape. Unlike a revolved solid, a revolved surface is open and does not enclose a volume. A revolve direction is also needed, as for a solid.

We will use the 2D Sketch tool to define the sketch shown in Fig. 10-14 in the front plane, which will be revolved about a vertical construction line or axis. The fillet radius is 1mm, as shown.

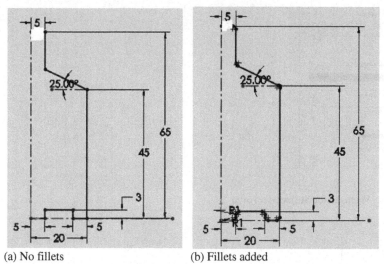

(a) No fillets   (b) Fillets added

Figure 10-14 Sketches for revolved surface

1. Create the sketch.
2. Click Insert > Surface > Revolve.
3. Select the contour of Sketch1 (Sketch1-Contour<1>) from the graphics window.
4. Select the direction, Line1, about which to revolve the contour (see Fig. 10-15).
5. Click OK.

(a) PropertyManager for Surface-Revolve

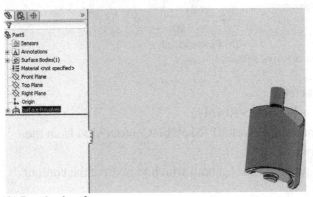

(b) Revolved surface
Figure 10-15 Revolved surface

10.8 Knitting Multiple Surfaces

We can join surfaces together using the Knit Surface tool. The surfaces are called knitting multiple surfaces bodies, just like we have solid bodies, which are multiple bodies. The surface bodies are *sewn together*, just like a tailor sewing pieces of cloth.

If the knit operation results in a closed surface, that is a watertight volume, the Fill Surface option will turn it into a solid.

10.9 Thicken Feature

If we select a closed surface body that encloses a volume, then there will be an option, Create solid from enclosed volume, and this will start the Thicken PropertyManager. This property manager can also be accessed through Insert > Boss/Base > Thicken.

Let us convert the modified freeform surface shown in Fig. 10-10 into a freeform solid. We will use the Thicken Feature option.
1. Click Insert > Boss/Base > Thicken.
2. Choose Boundary-Surface1 in the Thicken Parameter dialog (see Fig. 10-16).
   The converted solid is shown in Fig. 10-17.

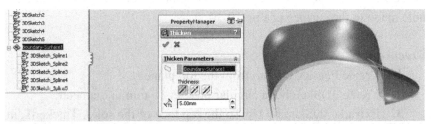

Figure 10-16 Process of converting a surface model into a solid model using the Thicken Feature

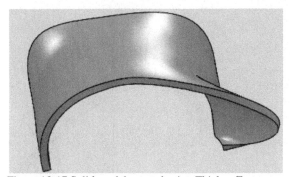

Figure 10-17 Solid model created using Thicken Feature

Once we have converted the surface model into a solid model using Thicken Feature, notice that two things happen:

1. The Surface Bodies(1), previously at the top of the FeatureManager, and the Boundary-Surface(1) that was at the bottom, have both disappeared.

2. The Thicken1 feature is added to the bottom of the FeatureManager.

These observations are important: we now have only solids, and so there are no references to surface bodies.

10.10 Fill Surface

The Fill Surface tool allows us to fill a surface to make it solid or to knit the surface into a surface body. We will demonstrate this tool by filling a hole made in the surface of Fig. 10-10.

Create a hole

1. Create Plane5, 60 mm from the top plane.

2. Create 3DSketch6, a circular profile, 15 mm in diameter.

3. Extrude 3DSketch6 Up To the top face of the solid, Thicken1, in the Direction of 3DSketch7, resulting in a hole (see Fig. 10-18).

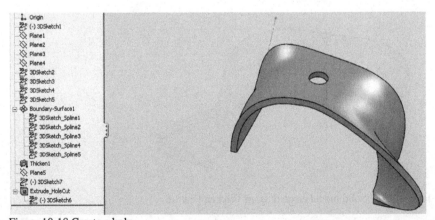

Figure 10-18 Create a hole

Fill the hole

4. Click Insert > Surface > Fill Surface (see Fig. 10-19 for preview).

5. Click the cut-out in the graphics window for the Patch Boundary rollout; this is the Extrude_HoleCut. Edge<1>Contact-50 appears in the Patch Boundary rollout and the cut is filled at the top (see Fig. 10-20). It is amazing. Notice that the solid is not filled; the top surface of the solid is filled. This is a similar principle to the "shut-off surfaces" of mold design (see Chapter 11).

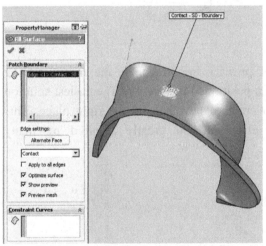

Figure 10-19 Preview for Fill Surface PropertyManager

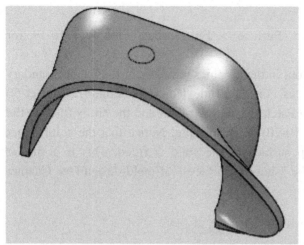

Figure 10-20 Repaired surface

The Fill Surface tool is of primary importance in the manufacturing industry. Say a hole has been positioned incorrectly. The easiest solution is to fill the hole and then create a new hole in the right place. This solution is effective and efficient, and greatly admired by many designers.

10.11   Extruded Surface: Type II

1. Click the top plane.
2. Create Sketch1 (see Fig. 10-21).
3. Click Insert > Surface > Extrude. The Surface-Extrude PropertyManager appears (see Fig. 10-22).
4. In the Direction 1 rollout, enter 35 mm for the extrusion depth.
5. Click OK. The extruded surface model is shown in Fig. 10-23.

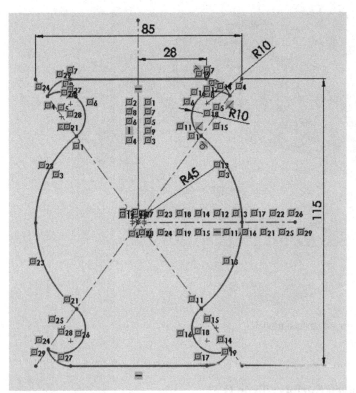

Figure 10-21 Sketch1

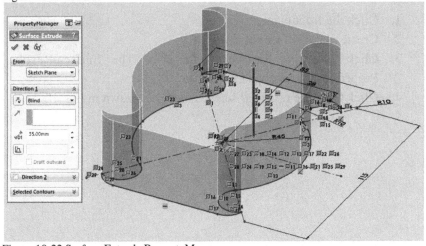

Figure 10-22 Surface-Extrude PropertyManager

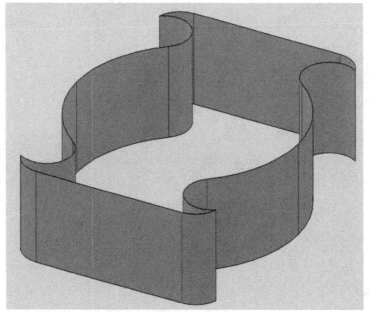

Figure 10-23 Extruded surface model

10.12   Revolved Surface: Type II

1. Click the front plane.
2. Create Sketch1 (see Fig. 10-24).
3. Click Insert > Surface > Revolve. The Surface-Revolve PropertyManager appears (see Fig. 10-25).
4. Click OK. The revolved surface model is shown in Fig. 10-26.

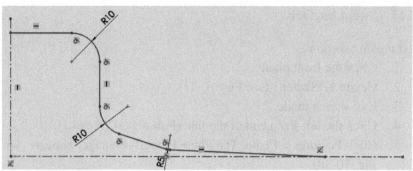

Figure 10-24 Sketch1

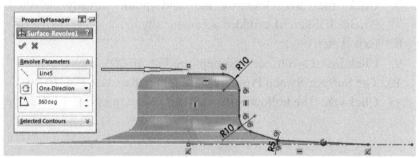

Figure 10-25 Surface-Revolve PropertyManager

Figure 10-26 Revolved surface model

10.13   Swept Surface

Rectangular section

1. Click the front plane.
2. Create 3DSketch1 (see Fig. 10-27).
3. Exit sketch mode.
4. Click the left-hand end of the line (Point9@3DSketch1).
5. Click Features > Plane. The Plane PropertyManager appears (see Fig. 10-28).
6. In the Selections rollout click Normal to Curve.
7. Click the arc close to the end point already selected. Arc3@3DSketch1 is added automatically.
8. Exit sketch mode.
9. Click Insert > Surface > Sweep (see Fig. 10-29).
10. The Surface-Sweep PropertyManager appears (see Fig. 10-30).
11. Click OK. The hollow surface model is shown in Fig. 10-31.

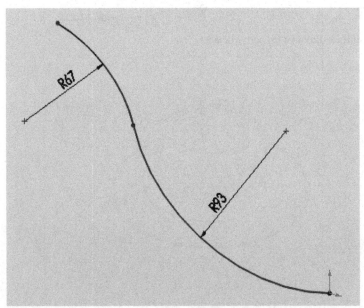

Figure 10-27 Sketch1

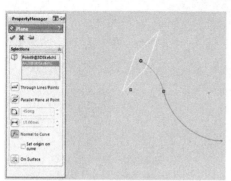
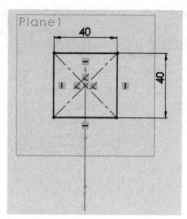

Figure 10-28 Plane PropertyManager

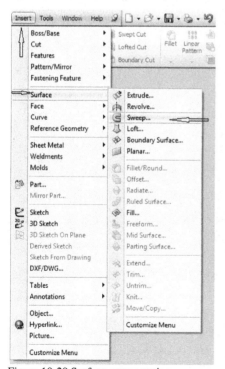

Figure 10-29 Surface sweep option

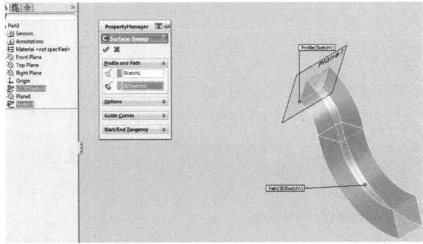

Figure 10-30 Surface-sweep PropertyManager

Figure 10-31 Surface model

Circular section

1. Click the front plane.
2. Create 3DSketch1 (see Fig. 10-32).
3. Exit sketch mode.
4. Click the right-hand end of the line (Point9@3DSketch1).
5. Click Features > Plane. The Plane PropertyManager appears (see Fig. 10-33(a)).
6. In the Selections rollout, click Normal to Curve.
7. Click the arc close to the end point already selected. Arc3@3DSketch1 is added automatically (see Fig. 10-33(b)).
8. Exit sketch mode.
9. Click Insert > Surface > Sweep (see Fig. 10-34).
10. The Surface-Sweep PropertyManager appears (see Fig. 10-35).
11. Click OK. The hollow swept surface model is shown in Fig. 10-36.

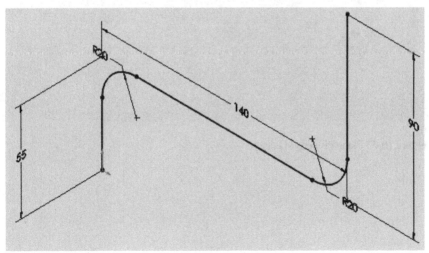

Figure 10-32 Sketch1

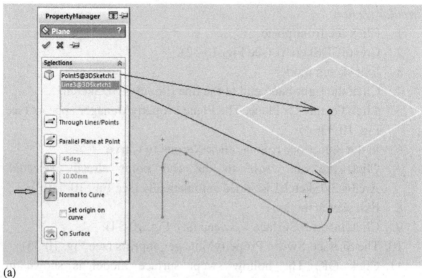

(a)

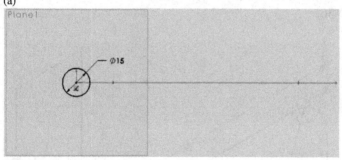

(b)
Figure 10-33 Plane PropertyManager

Surface Modeling

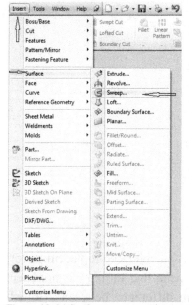

Figure 10-34 Surface sweep option

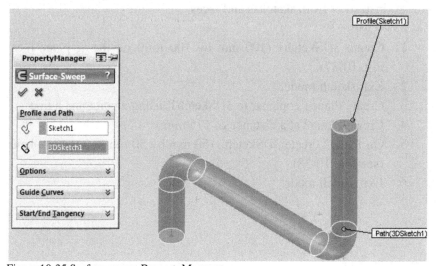

Figure 10-35 Surface-sweep PropertyManager

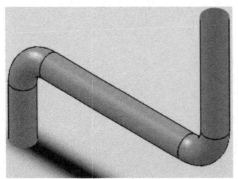

Figure 10-36 Swept surface model

10.14 Lofted Surface

The boundary surface option was created as an enhanced replacement for the lofted surface, but certain limitations mean that lofting has not been removed from the features list. The main difference between a lofted surface and a lofted solid is that a surface can be lofted using edges and curves, instead of from sketches and faces.

1. Create 3DSketch1 (100 mm by 100 mm) on the yz-plane (see Fig. 10-37).
2. Exit sketch mode.
3. Create Plane1 coplanar to 3DSketch1, using a point and a line.
4. Create Plane2 at a distance of 175 mm.
5. On Plane2 create 3DSketch2 (50 mm by 50 mm) on the yz-plane (see Fig. 10-38).
6. Exit sketch mode.

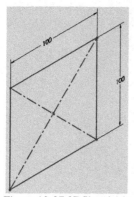

Figure 10-37 3DSketch1 in the *yz*-plane

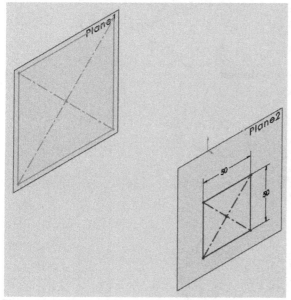

Figure 10-38 3DSketch2 on *yz*-plane of Plane2

7. Click Insert > Surface > Loft. The Surface-Loft Property-Manager appears (see Fig. 10-39).

8. Click OK. The hollow lofted surface model is shown in Fig. 10-40.

431

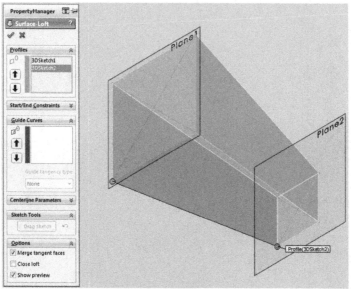

Figure 10-39 Surface-Loft PropertyManager

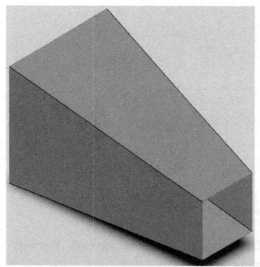

Figure 10-40 Lofted surface model

10.15   Freeform Surface Design: Boundary Surface

Control polygons

1. Start a new SolidWorks part document.
2. Select the front plane.
3. Click Sketch 3DSketch.
4. Create 3DSketch1 on the *xy*-plane (see Fig. 10-41).
5. Exit sketch mode.
6. Create Plane1 coplanar to 3DSketch1, using a point and a line.
7. Create three planes parallel to 3DSketch1: Plane2, Plane3, and Plane4 (see Fig. 10-42).

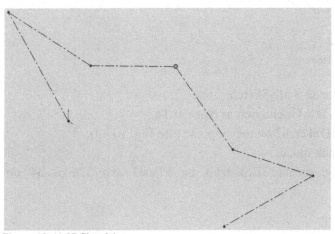

Figure 10-41 3DSketch1

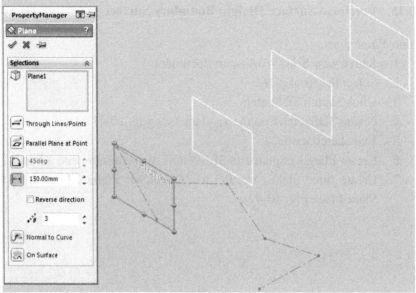

Figure 10-42 Four planes

8. Click Plane2 > 3D-Sketch.
9. Set the View Orientation as Normal To.
10. Create 3DSketch2 on the *xy*-plane (see Fig. 10-43).
11. Exit sketch mode.
12. Similarly, create 3DSketch3 on Plane3 and 3DSketch4 on Plane4.

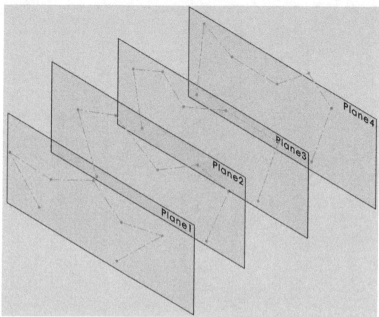

Figure 10-43 Four control polygons

Modifying the control polygons

    13. Click Plane1 > 3D-Sketch.

    14. Modify vertices as desired (see Fig. 10-44 for modified vertices in Plane1 and Plane4).

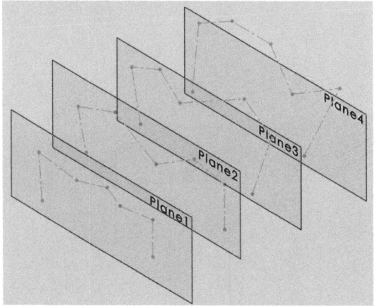

Figure 10-44 Modified control polygons

U-direction control curves

15. Click Insert > Curve > Curve Through Reference Points (see Fig. 10-45).

16. The Curve Through PropertyManager automatically appears (see Fig. 10-46).

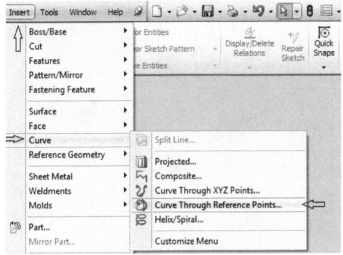

Figure 10-45 Start the spline Curve Through Reference Points tool

17. Select vertices for the first set of control vertices (see Fig. 10-46).

18. Click OK.

19. Do the same for the second, third, and fourth sets of control vertices (see Figs 10-47 to 10-49).

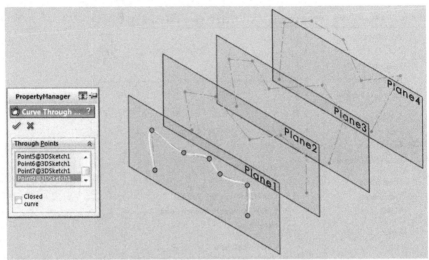

Figure 10-46 Spline Curve Through Reference Points PropertyManager for first U-points

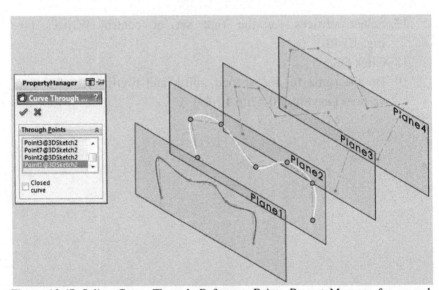

Figure 10-47 Spline Curve Through Reference Points PropertyManager for second U-points

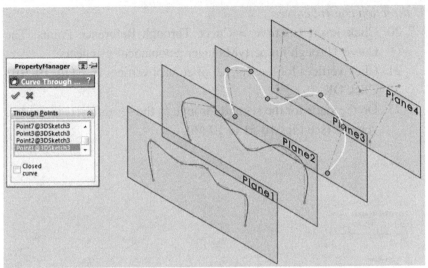

Figure 10-48 Spline Curve Through Reference Points PropertyManager for third U-points

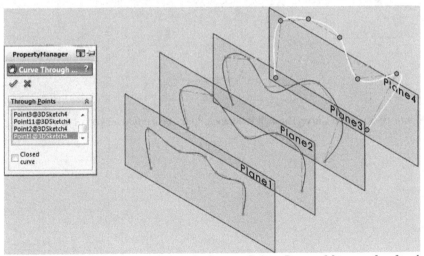

Figure 10-49 Spline Curve Through Reference Points PropertyManager for fourth U-points

U-direction control curves

20. Click Insert > Curve > Curve Through Reference Points. The Curve Through PropertyManager automatically appears.

21. Click vertices for the first set of control vertices (see Fig. 10-50)

22. Click OK.

23. Do the same for the second through to the seventh sets of control vertices (see Figs 10-51 to 10-56).

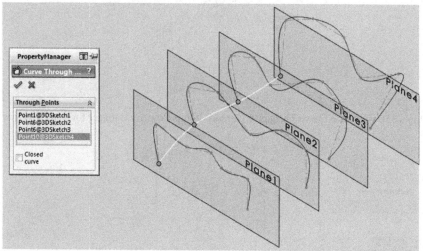

Figure 10-50 Spline Curve Through Reference Points PropertyManager for first V-points

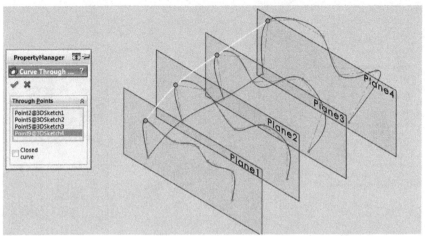

Figure 10-51 Spline Curve Through Reference Points PropertyManager for second V-points

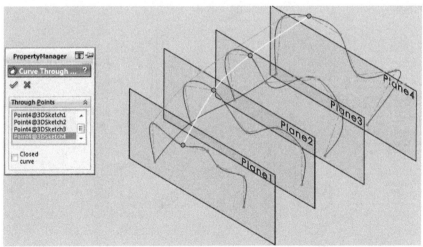

Figure 10-52 Spline Curve Through Reference Points PropertyManager for third V-points

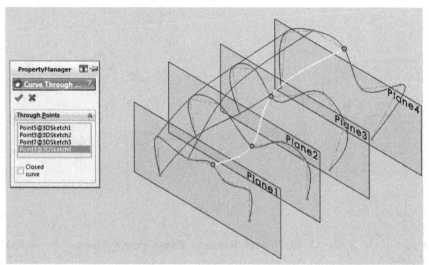

Figure 10-53 Spline Curve Through Reference Points PropertyManager for fourth V-points

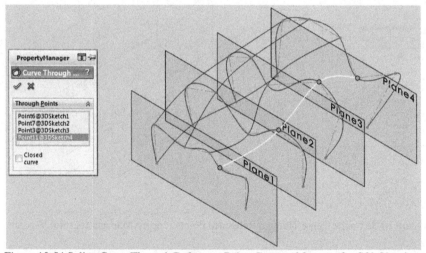

Figure 10-54 Spline Curve Through Reference Points PropertyManager for fifth V-points

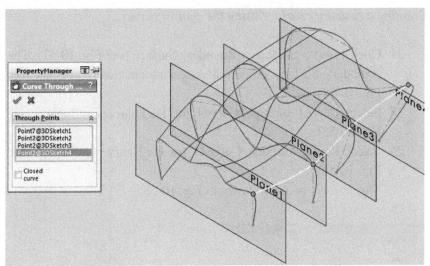

Figure 10-55 Spline Curve Through Reference Points PropertyManager for sixth V-points

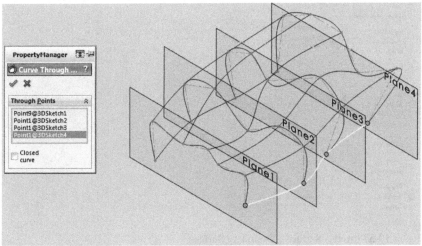

Figure 10-56 Spline Curve Through Reference Points PropertyManager for seventh V-points

Creating a boundary surface using the control curves

24. Click Insert > Surface > Boundary Surface (see Fig. 10-57). The Boundary Surface PropertyManager automatically appears (see Fig. 10-58).

25. Select Curve1 through Curve4 for Dir1 curves influence (see Fig. 10-58).

26. Select Curve5 through Curve11 for Dir2 curves influence (see Fig. 10-58).

The surface model is shown in Fig. 10-58.

Figure 10-57 Insert > Surface > Boundary Surface

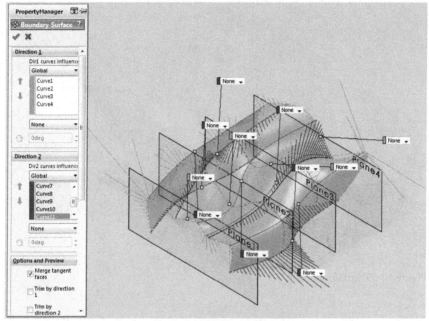

Figure 10-58 Boundary Surface PropertyManager

The surface model is shown in Fig. 10-59. By modifying the control vertices, a modified surface model can be created, as shown in Fig. 10-60. The FeatureManager for the surface is shown in Fig. 10-61.

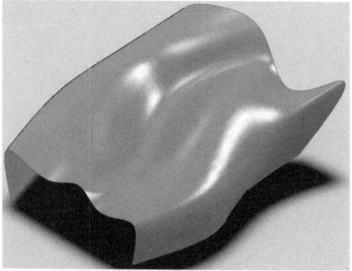

Figure 10-59 Surface model

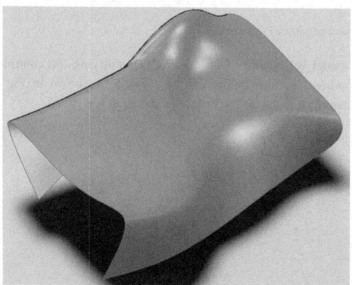

Figure 10-60 Surface model created by modifying some control vertices

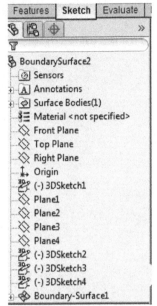

Figure 10-61 FeatureManager for the surface

Summary

This chapter has presented three methods of creating a solid part using surfaces. Try all of these methods. After all, a solid is made up of the surfaces that define its boundary, therefore, in general it is natural to piece together surfaces into a solid. On the other hand, the outer parts of a solid can be extracted to give bounding surfaces. We have also presented a very useful tool for filling holes in a surface, which is widely used in manufacturing.

Exercises

1. Create (drill) a hole, 10 mm diameter anywhere on top of Fig. 10-59, then fill the hole.
2. Repeat the above process for Fig. 10-60.
3. How is the process of filling a hole useful in manufacturing?

Chapter 11

Mold Design

Objectives:

When you complete this chapter you will:

- Understand the terminology and concepts of draft, parting line, draft analysis, shut-off surfaces, parting surfaces, and tooling split used in mold design
- Have learnt how to create a parting line
- Have learnt how to create a parting surface
- Have learnt how to carry out a tooling split and create cavity and core blocks for injection molds

11.1 Mold Design Background

Mold design is a specialized area of manufacturing, which creates the designs for the lower-half and the upper-half of molds required for producing forged, cast, and plastic parts. Very little information is available about any proven methodologies of using automated software. Mold design is one of the important applications of SolidWorks. Earlier versions, such as the 2009 version, followed a long-winded approach, but SolidWorks 2009 is quite effective in handling many different mold designs. The mold tools help users create cavity and core blocks for injection molds. However, libraries have not been provided for building the entire mold or mold components.

Creating a parting surface in SolidWorks works best if the planar parting lines are convex throughout the part. In creating a mold design, it is common to spend 70% or more of the time on the parting surface. SolidWorks mold tools do not work well if there are concave parting lines/surfaces or non-planar lines/surfaces, which are commonly encountered in complex parts. SolidWorks mold tools are only semi-automated. Creating a moderately complex mold requires some level of manual intervention to get a reasonably usable result. Experienced mold designers tend to use techniques such as cutting away pieces of a solid, surface modeling, and manual surfacing.

The remainder of this chapter presents the application of SolidWorks to mold design. The prerequisite is that the part should have been already modeled. Let us now follow, step-by-step, the method used in SolidWorks mold tools for designing a mold. We will use parts that we have already modeled: basin, pulley (6.1.1), bowl, plastic cover (5.3.2), and sump.

11.2 Overview of Mold Design Tools

Mold analysis tools are used by designers of molded plastic parts and by designers of the mold tools used to manufacture those parts. A mold is created using a sequence of integrated tools that control the process. These mold tools can be used to analyze and correct deficiencies with either SolidWorks models or imported models. The tools help with the initial analysis through to the creation of the tooling split. The result of the tooling split is a multi-body part containing separate bodies for the molded part, the core, and the cavity, plus other optional bodies such as side cores. The multi-body part file maintains the whole design in one convenient location. Changes to a molded part are automatically reflected in the tooling bodies.

SolidWorks have provided an overview, listed in Table 11-1, of the typical mold design tasks and their solutions.

Mold Design

Table 11-1 Mold design overview from SolidWorks

Tasks	Solutions
When you are not using models built with SolidWorks, import parts into SolidWorks.	Use the Import/Export tools to import models into SolidWorks from another application. The model geometry in imported parts can include imperfections such as gaps between surfaces. The SolidWorks application includes an import diagnostic tool to address these issues.
Determine if a model (imported or built in SolidWorks) includes faces without draft.	Use the Draft Analysis tool to examine the faces to ensure sufficient draft. Additional functionality includes: **Face classification:** Display color-coded count of faces with positive draft, faces with negative draft, faces with insufficient draft, and straddle faces. **Gradual transition:** Display the draft angle as it changes within each face.
Check for undercut areas.	Use the Undercut Detection tool to locate trapped areas in a model that prevent ejection from the mold. These areas require a mechanism called a "side core" to produce the undercut relief. Side cores eject from the mold as it is opened.
Scale the model.	Resize the model's geometry with the Scale tool to account for the shrink factor when plastic cools. For odd-shaped parts and glass-filled plastic, you can specify non-linear values.
Select the parting lines from which you create the parting surface.	Generate parting lines with the Parting Lines tool that selects a preferred parting line around the model.

Table 1 Mold Design overview from SolidWorks: contd.

Create shut-off surfaces to prevent leakage between core and cavity.	Detect possible sets of holes and automatically shut them off with the Shut-off Surfaces tool. The tool creates surfaces to fill the open holes using no fill, tangent fill, contact fill, or a combination of the three. The no fill option is used to exclude one or more through holes so you can manually create their shut-off surfaces. You can then create the core and cavity.
Create the parting surface, from which you can create the tooling split. With certain models, use the ruled surface tool to create interlock surfaces along the edges of the parting surface.	Use the Parting Surface tool to extrude surfaces from the parting lines generated earlier. These surfaces are used to separate the mold cavity geometry from the mold core geometry.
Add interlock surfaces to the model.	Apply these solutions for interlock surfaces: **Simpler models:** Use the automated option that is part of the Tooling Split tool. **More complex models:** Use the Ruled Surface tool to create the interlock surfaces.
Perform tooling split to separate core and cavity.	Create the core and cavity automatically with the Tooling Split tool. The Tooling Split tool uses the parting line, shut-off surfaces, and parting surfaces information to create the core and cavity, and allows you to specify the block sizes.
Create side cores, lifters, and trimmed ejector pins.	Use Core to extract geometry from the tooling solid to create a core feature. You can also create lifters and trimmed ejector pins.
Display the core and cavity transparently, enabling you to view the model inside.	Assign different colors to each entity with the Edit Color tool. The Edit Color tool also manipulates optical properties such as transparency.
Display the core and cavity separated.	Separate the core and cavity at a specified distance with the Move/Copy Bodies tool.

11.3   Mold Design Methodology

The general method of a creating a mold design is as follows:
1. Click Insert > Part > Name*.sldprt.
2. Click Draft > DraftXpert > Add. Check Auto paint.
3. Click Scale and enter 1.2. Check Uniform Scaling.
4. Click Parting Line, and under Mold Parameters enter the pull direction. Use View > Temporary Axes and note the direction. Click Draft Analysis. Check Use for Core/Cavity Split and click OK. Pick all lines/curves defining the parting line and click OK. An arrow will move along the parting line contour. It should be manually guided.
5. Click Shut-Off Surfaces and click OK. Clear the Knit option if advised. If redundant shut-off surfaces exist, delete them.
6. Click Parting Surface > Top View > Value <1-50>. Check Perpendicular to pull. If there is a warning, check Knit all surfaces.
7. Click the top plane (or another plane) and the sketch base (normally rectangular) of the tooling split.
8. Exit sketch mode.
9. Check Tooling Split > plane in Step 6 and define Heights Up/Down from this plane, which passes through the origin. Define the height of the tooling split upward from the datum. Define the height of the tooling split downward from the datum.
10. Right-click the surface bodies and hide them.
11. Expand the solid bodies in the FeatureManager.
12. Click Tooling Split[1] > Insert > FeatureManager > Move-Copy. Move the vertical axis of the triad in the direction to pull out.
13. Repeat Step 12 for Tooling Split[2].

11.4   Basin

1. Open a SolidWorks part document. Mold Tool should be activated.
2. Set the document properties for the model, with decimal places equal to 2.
3. Click Insert > Part > BASIN*.sldprt and click OK. This inserts the part at the origin (see Fig. 11-1).
4. Click Draft > DraftXpert > Add. Check Auto paint and click OK (see Fig. 11-2).

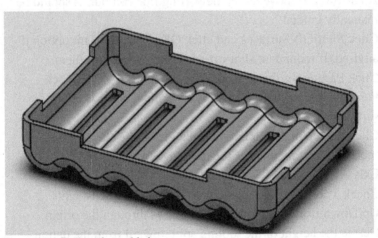

Figure 11-1 Basin to be molded

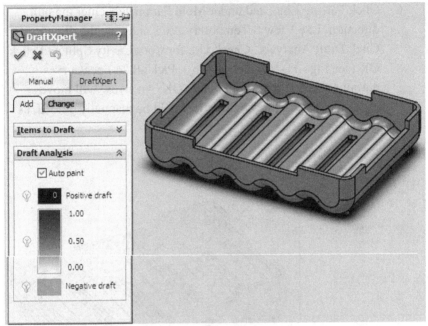

Figure 11-2 DraftXpert

5. Click Scale and enter 1.2. Check Uniform Scaling (see Fig. 11-3).

6. Click Parting Line, and under Mold Parameters enter the pull direction. Use View > Temporary Axes and note the direction. Click Draft Analysis. Check Use for Core/Cavity Split and click OK (see Figs 11-4(a) and 11-4(b)). Pick all 24 lines/curves defining the parting line and click OK. An arrow will move along the parting line contour. It should be manually guided, as shown in Fig. 11-5.

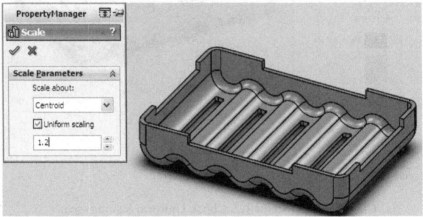

Figure 11-3 Scaling

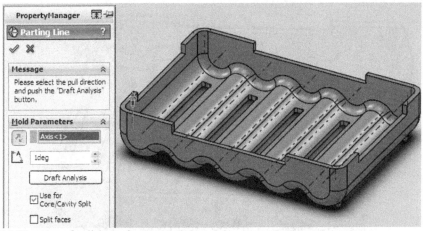

(a) Direction of pull (arrow)

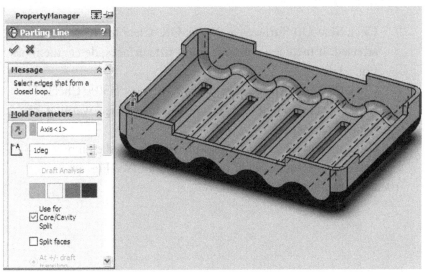

(b) Draft analysis

Figure 11-4 Pulling direction and draft analysis for parting line

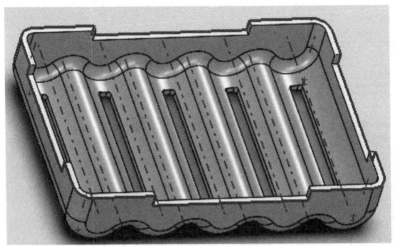

Figure 11-5 Parting line

7. Click Shut-Off Surfaces and click OK. Clear the Knit option if advised. If there are redundant shut-off surfaces, delete them (see Fig. 11-6).

8. Click Parting Surface > Top View > Value <1-50>. Check Perpendicular to pull (see Fig. 11-7). If there is a warning, check Knit all surfaces.

9. Click the top plane (or another plane) and the sketch base (normally rectangular) of the tooling split (see Fig. 11-8).

10. Exit sketch mode.

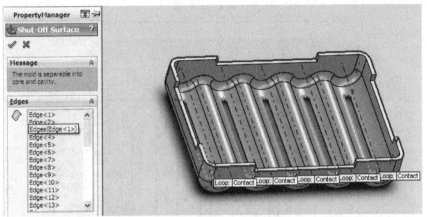

Figure 11-6 Shut-off surfaces

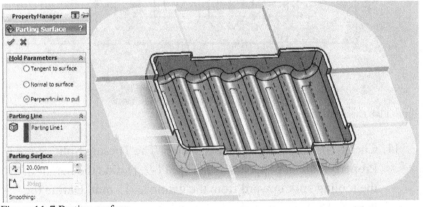

Figure 11-7 Parting surface

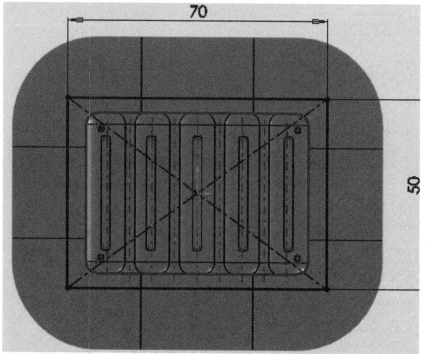

Figure 11-8 Base of tooling split

11. Check Tooling Split > plane in step 6. The Tooling Split PropertyManager appears as in Fig. 11-9. Define the height of the tooling split upward from the datum as 15 mm. Define the height of the tooling split downward from the datum as 5 mm. Click OK and hide the parting surface as well as the temporary axes (see Fig. 11-10).

12. Right-click the surface bodies and hide them.

13. Expand the solid bodies in the FeatureManager

14. Click Tooling Split[1] > Insert > Feature > Move-Copy. Move the vertical axis of the triad in the direction to pull out (see Fig. 11-11).

15. Repeat Step 14 for Tooling Split[2] (see Fig. 11-12).

The mold design process is complete and Fig. 11-13 shows the orientations of the top and bottom of the mold. These will be needed by the manufacturing department. Figure 11-14 summarizes the mold design process.

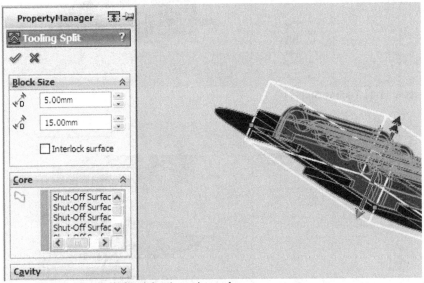

Figure 11-9 Tooling split for sizing lower/upper boxes

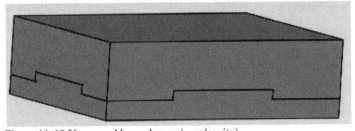

Figure 11-10 Upper and lower boxes (core/cavity)

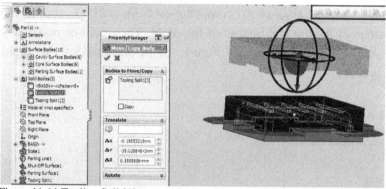

Figure 11-11 Tooling Split[1]

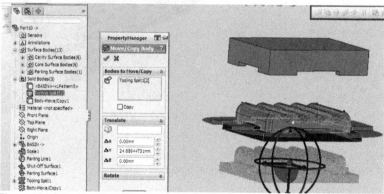

Figure 11-12 Tooling Split[2]

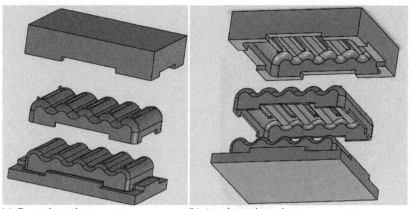

(a) One orientation   (b) Another orientation

Figure 11-13 Orientations of part and molds

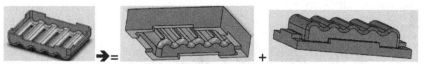

Figure 11-14 Summary of the mold design process

11.5 Pulley

1. Open a SolidWorks part document. Mold Tool should be activated.
2. Set the document properties for the model, with decimal places equal to 2.
3. Click Insert > Part > PULLEY*.sldprt and click OK. This inserts the part at the origin (see Fig. 11-15).
4. Click Draft > DraftXpert > Add. Check Auto paint and click OK (see Fig. 11-16).

Figure 11-15 Pulley to be molded

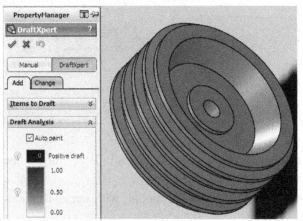

Figure 11-16 DraftXpert

5. Click Scale and enter 1.2. Check Uniform Scaling (see Fig. 11-17).

6. Click Parting Line, and under Mold Parameters enter the pull direction. Use View > Temporary Axes and note the direction. Click Draft Analysis. Check Use for Core/Cavity Split and click OK (see Figs 11-18(a) and 11-18(b)). Pick all 24 lines/curves defining the parting line and click OK. An arrow will move along the parting line contour. It should be manually guided, as shown in Fig. 11-19.

464

Figure 11-17 Scaling

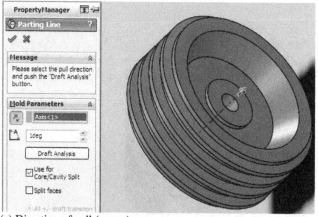

(a) Direction of pull (arrow)

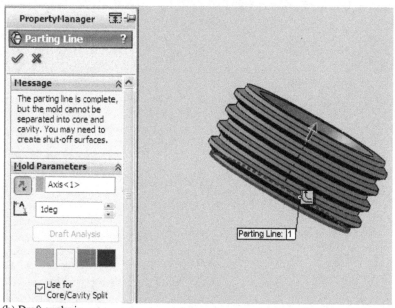

(b) Draft analysis

Figure 11-18 Direction of pull and draft analysis for parting line

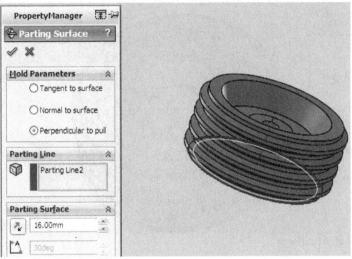

Figure 11-19 Parting Line

7. Click Shut-Off Surfaces and click OK. Clear Knit option if advised. If there are redundant shut-off surfaces, delete them (see Fig. 11-20).

8. Click Parting Surface > Top View > Value <1-50>. Check Perpendicular to pull (see Fig. 11-21). If there is a warning, check Knit all surfaces.

9. Click the top plane (or another plane) and the sketch base (normally rectangular) of the tooling split (see Fig. 11-22).

10. Exit sketch mode.

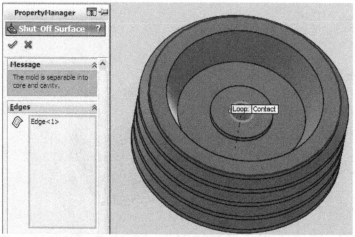

Figure 11-20 Shut-off surfaces

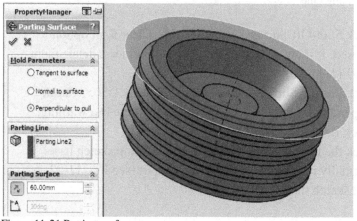

Figure 11-21 Parting surface

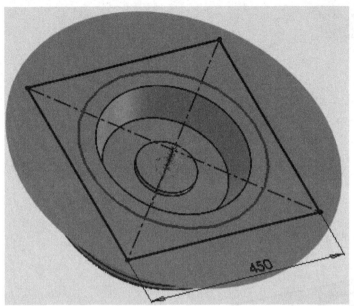

Figure 11-22 Base of tooling split

11. Check Tooling Split > plane in step 6. The Tooling Split
    PropertyManager appears, as in Fig. 11-23. Define the height of
    the tooling split upward from the datum as 15 mm. Define the
    height of the tooling split downward from the datum as 5 mm.
    Click OK and hide the parting surface as well as the temporary
    axes (see Fig. 11-24).
12. Right-click the surface bodies and hide them.
13. Expand the solid bodies in the FeatureManager.
14. Click Tooling Split[1] > Insert > Feature > Move-Copy.
    Move the vertical axis of the triad in the direction to pull out
    (see Fig. 11-25).
15. Repeat Step 14 for Tooling Split[2] (see Fig. 11-26).

The mold design process is complete and Fig. 11-27 shows the
orientations of the top and bottom of the mold. Fig. 11-28 summarizes
the mold design process.

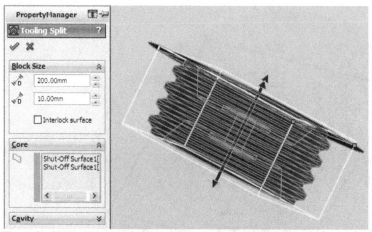

Figure 11-23 Tooling split for sizing lower/upper boxes

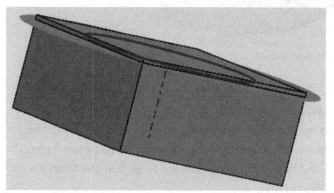

Figure 11-24 Upper and lower boxes (core/cavity)

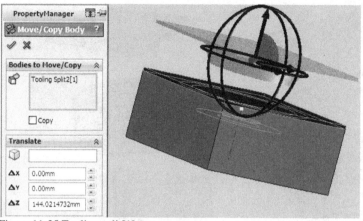

Figure 11-25 Tooling split[1]

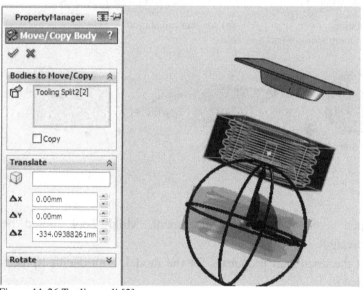

Figure 11-26 Tooling split[2]

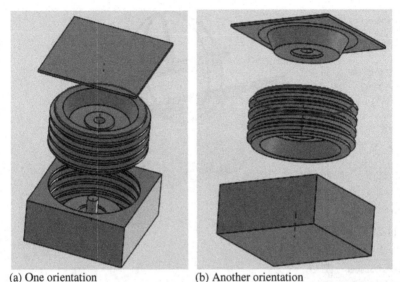

(a) One orientation   (b) Another orientation

Figure 11-27 Orientations of part and molds

Figure 11-28 Summary of the mold design process

11.6 Bowl

1. Open a SolidWorks part document. Mold Tool should be activated.
2. Set the document properties for the model, with decimal places equal to 2.
3. Click Insert > Part > BOWL*.sldprt and click OK. This inserts the part at the origin (see Fig. 11-29).
4. Click Draft > DraftXpert > Add. Check Auto paint and click OK (see Fig. 11-30).

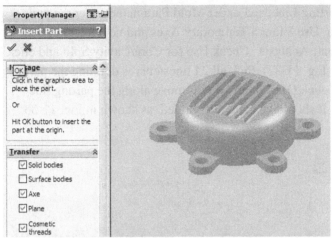

Figure 11-29 Bowl to be molded

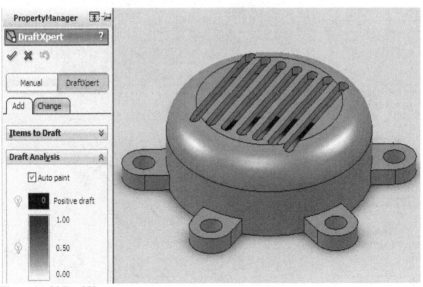

Figure 11-30 DraftXpert

5. Click Scale and enter 1.2. Check Uniform Scaling (see Fig. 11-31).

6. Click Parting Line, and under Mold Parameters enter the pull direction. Use View > Temporary Axes and note the direction. Click Draft Analysis. Check Use for Core/Cavity Split and click OK (see Fig. 11-32). Pick all 24 lines/curves defining the parting line and click OK. An arrow will move along the parting line contour. It should be manually guided, as shown in Fig. 11-33.

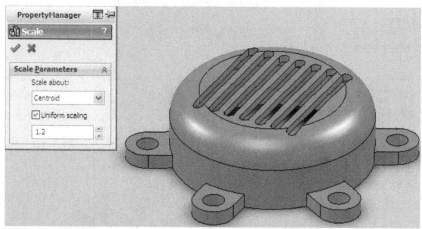

Figure 11-31 Scaling

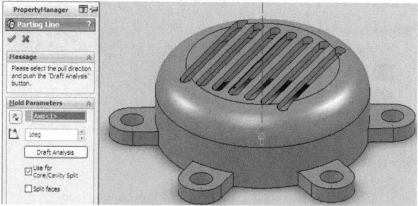

Figure 11-32 Direction of pull (arrow)

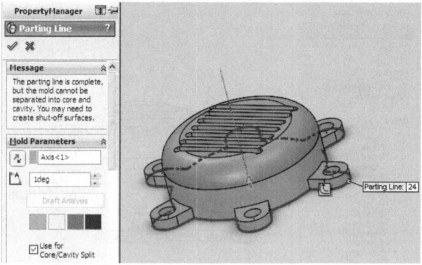

Figure 11-33 Parting Line

7. Click Shut-Off Surfaces and click OK. Clear Knit option if advised. If redundant shut-off surfaces exist, delete them (see Fig. 11-34).

8. Click Parting Surface > Top View > Value <1-50>. Check Perpendicular to pull (see Fig. 11-35). If there is a warning, check Knit all surfaces.

9. Click the top plane (or another plane) and the sketch base (normally rectangular) of the tooling split (see Fig. 11-36).

10. Exit sketch mode.

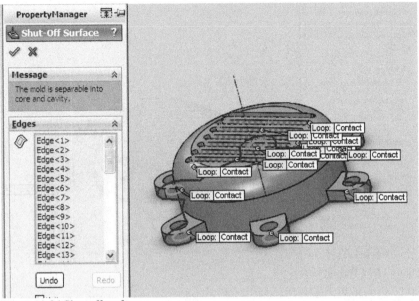

Figure 11-34 Shut-off surfaces

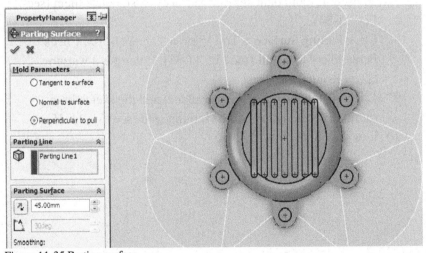

Figure 11-35 Parting surface

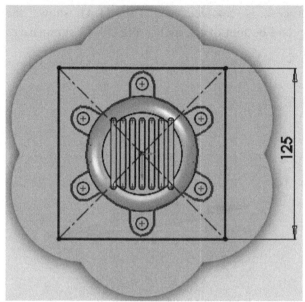

Figure 11-36 Base of tooling split

11. Check Tooling split > plane in step 6. The Tooling Split PropertyManager appears, as in Fig. 11-37. Define the height of the tooling split upward from the datum as 15 mm. Define the height of the tooling split downward from the datum as 5 mm. Click OK and hide the parting surface as well as the temporary axes (see Fig. 11-38).

12. Right-click the surface bodies and hide them.

13. Expand the solid bodies in the FeatureManager.

14. Click Tooling Split[1] > Insert > Feature > Move-Copy. Move the vertical axis of the triad in the direction to pull out (see Fig. 11-39).

15. Repeat Step 14 for Tooling Split[2] (see Fig. 11-40).

The mold design process is complete and Fig. 11-41 shows the orientations of the top and bottom of the mold. Fig. 11-42 summarizes the mold design process.

Figure 11-37 Tooling split for sizing lower/upper boxes

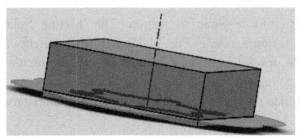

Figure 11-38 Upper and lower boxes (core/cavity)

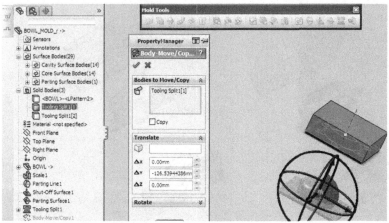

Figure 11-39 Tooling split[1]

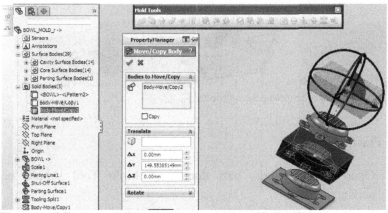

Figure 11-40 Tooling split[2]

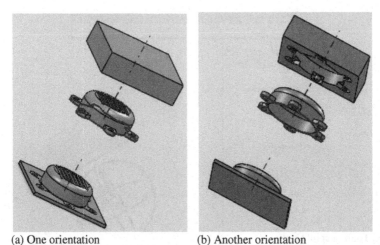

(a) One orientation   (b) Another orientation

Figure 11-41 Orientations of part and molds

Figure 11-42 Summary the mold design process

11.7 Plastic Cover

1. Open a SolidWorks part document. Mold Tool should be activated.
2. Set the document properties for the model, with decimal places equal to 2.
3. Click Insert > Part > PLASTIC_COVER*.sldprt and click OK. This inserts the part at the origin (see Fig. 11-43).

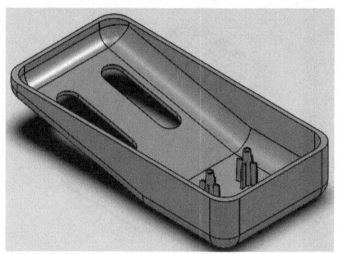

Figure 11-43 Plastic cover to be molded.

4. Click Scale and enter 1.2. Check Uniform Scaling.
5. Click Parting Line, and under Mold Parameters enter the pull direction. Use View > Temporary Axes and note the direction. Click Draft Analysis. Check Use for Core/Cavity Split and click OK (see Fig. 11-44). Pick all 24 lines/curves defining the parting line and click OK. An arrow will move along the parting line contour. It should be manually guided. Fig. 11-45 shows the parting line.

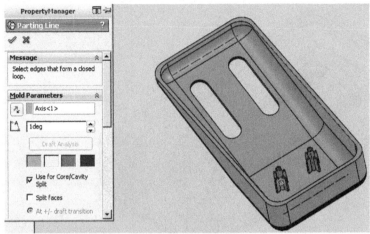

Figure 11-44 Draft analysis

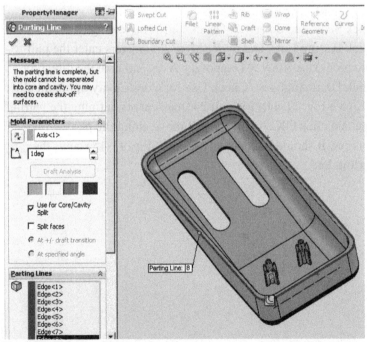

Figure 11-45 Parting Line

6. Click Shut-Off Surfaces and click OK. Clear Knit option if advised. If redundant shut-off surfaces exist, delete them (see Fig. 11-46).

7. Click Parting Surface > Top View > Value <1-50>. Check Perpendicular to pull (see Fig. 11-47). If there is a warning, check Knit all surfaces.

8. Click the top plane (or another plane) and the sketch base (normally rectangular) of the tooling split (see Fig. 11-48).

9. Exit sketch mode.

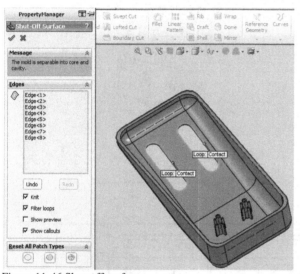

Figure 11-46 Shut-off surfaces

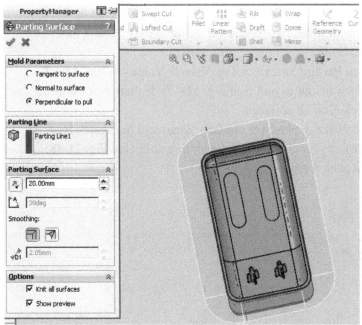

Figure 11-47 Parting surface

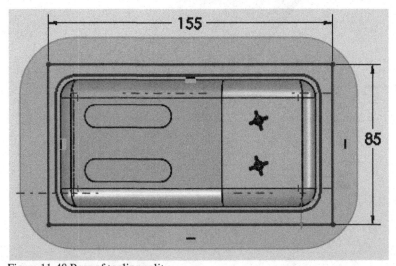

Figure 11-48 Base of tooling split

10. Check Tooling split > plane in step 6. The Tooling Split PropertyManager appears as in Fig. 11-49. Define the height of the tooling split upward from the datum as 15 mm. Define the height of the tooling split downward from the datum as 5 mm. Click OK and hide the parting surface as well as the temporary axes.

11. Right-click the surface bodies and hide them.

12. Expand the solid bodies in the FeatureManager.

13. Click Tooling Split[1] > Insert > Feature > Move-Copy. Move the vertical axis of the triad in the direction to pull out (see Fig. 11-50).

14. Repeat Step 14 for Tooling Split[2] (see Fig. 11-51).

The mold design process is complete and Fig. 11-52 shows the orientations of the top and bottom of the mold.

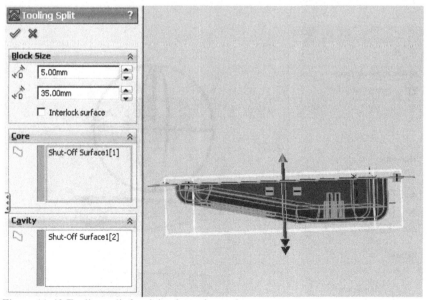

Figure 11-49 Tooling split for sizing lower/upper boxes

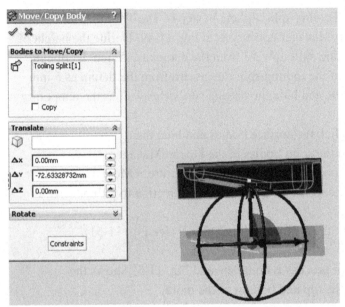

Figure 11-50 Tooling split[1]

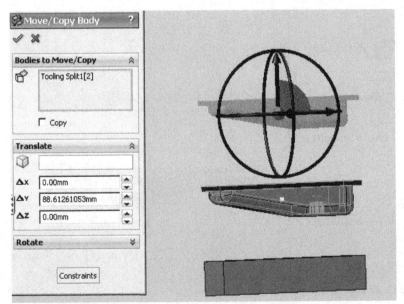

Figure 11-51 Tooling split[2]

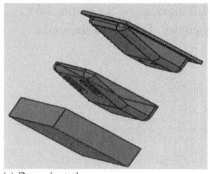
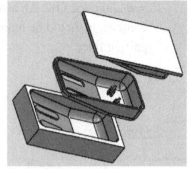

(a) One orientation   (b) Another orientation

Figure 11-52 Orientations of part and molds

11.8   Sump

1. Open a SolidWorks part document. Mold Tool should be activated.
2. Set the document properties for the model, with decimal places equal to 2.
3. Click Insert > Part > SUMP*.sldprt and click OK. This inserts the part at the origin (see Fig. 11-53).

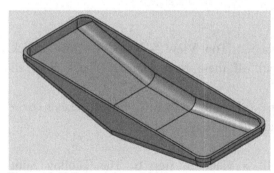

Figure 11-53 Sump to be molded

4. Click Scale and enter 1.2. Check Uniform Scaling.
5. Click Parting Line, and under Mold Parameters enter the pull direction. Use View > Temporary Axes and note the direction. Click Draft Analysis. Check Use for Core/Cavity Split and click OK (see Fig. 11-44). Pick all 24 lines/curves defining the parting

line and click OK. An arrow will move along the parting line contour. It should be manually guided. Fig. 11-54 shows the parting line.

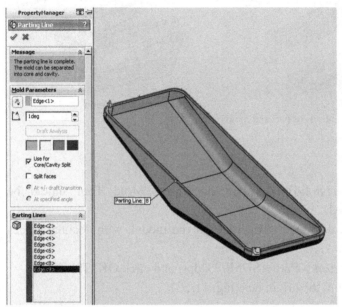

Figure 11-54 Parting Line

6. Click Parting Surface > Top View > Value <1-50>. Check Perpendicular to pull. If there is a warning, check Knit all surfaces.

7. Click the top plane (or another plane) and the sketch base (normally rectangular) of the tooling split.

8. Exit sketch mode.

9. Check Tooling split > plane in step 6. The Tooling Split Manager appears as in Fig. 11-55. Define the height of the tooling split upward from the datum as 15 mm. Define the height of the tooling split downward from the datum as 5 mm. Click OK and hide the parting surface as well as temporary axes.

10. Right-click the surface bodies and hide them.

11. Expand the solid bodies in the FeatureManager.

12. Click Tooling Split[1] > Insert > Feature > Move-Copy. Move the vertical axis of the triad in the direction to pull out.

13. Repeat Step 12 for Tooling Split[2].

The mold design process is complete as shown in Fig. 11-56.

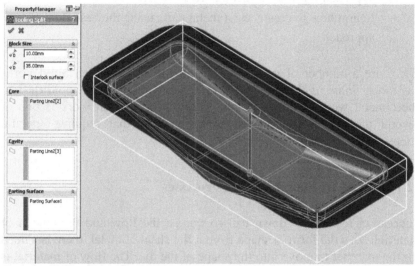

Figure 11-55 Base of tooling split

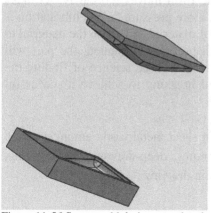

Figure 11-56 Sump mold design completed

Chapter 12
Sheet Metal Part Design

Objectives:

When you complete this chapter you will have:

- Learnt the differences between insert bends and base flange approaches for sheet metal parts
- Learnt how to create sheet metal parts using the base flange approach

There are a number of manufacturing functions for which SolidWorks has tools. Some of these manufacturing functions include mold design, sheet metal work, weldments, hinges, etc. Sheet metal work is one of several aspects of the manufacturing process that requires specific attention.

12.1 Sheet Metal Manufacturing Processes

Success in forming is down to two things: the flow and the stretch of material. As a die forms a shape from a flat sheet of metal, there is a need for the material to move into the shape of the die. The flow of material is controlled through pressure applied to the blank, aided by lubrication applied to the die or the blank. If the form moves too easily, wrinkles will occur in the part. To correct this, more pressure or less lubrication is applied to the blank to limit the flow of material and cause the material to stretch and become thin. If too much pressure is applied, the part will become too thin and break. Drawing metal is the science of finding the correct balance between wrinkles and breaking to achieve a successful part.

There are several methods involved in sheet metal work, amongst which are the following: bending, roll-forming, deep-drawing, bar-drawing, tube-drawing, wire-drawing, and plastic-drawing.

Bending

Bending is a common metalworking technique of processing sheet metal, as shown in Fig. 12-1. It is usually done by hand on a box and pan brake, or industrially on a brake press or machine brake. Typical products that are made like this are boxes, such as electrical enclosures and rectangular ductwork. Usually bending has to overcome both tensile stresses as well as compressive stresses. When bending is carried out, the residual stresses make the sheet spring back towards its original position, so we have to over-bend the sheet, keeping in mind the residual stresses. When sheet metal is bent, it stretches in length. The bend deduction is the amount the sheet metal will stretch when bent as measured from the outside. A bend has a radius. The term bend radius refers to the inside radius. The bend radius depends upon the dies used, the metal properties, and the metal thickness.

For sheet metal parts, the following are important:
- Bending phenomenon (effect of bend radius, material thickness, and bend angle)
- Stress relief methods (rectangular, tear, and obround configurations)

Many software packages refer to the K-factor for bending sheet metal. The K-factor is a ratio that represents the location of the neutral sheet with respect to the inside thickness of the sheet metal part. The bend allowance is the length of the arc of the neutral axis between the tangent points of a bend in any material.

$$B_d = 2 \times (R+T) - B_a$$
$$B_a = \pi \times (R + K*T) \times \alpha / 180$$
$$K = (180 \times B_a)(\pi \times \alpha \times T) - R/T$$

where

- B_a is the bend allowance
- R is the inside bend radius
- K is the K-factor, which is t / T
- T is the material thickness
- t is the distance from the inside face to the neutral sheet

- A is the bend angle in degrees (the angle through which the material is bent)

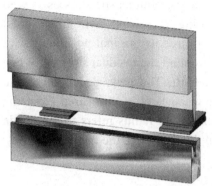

Figure 12-1 Sheet metal bending setup

Rolling-forming

Roll-forming is a continuous bending operation in which a long strip of metal (typically coiled steel) is passed through consecutive sets of rolls, or *stands*, each performing only an incremental part of the bend, until the desired cross-section profile is obtained (see Fig. 12-2). Roll-forming is ideal for producing parts with long lengths or in large quantities.

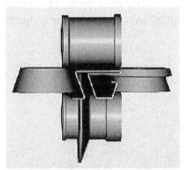

Figure 12-2 Sheet metal roll-forming setup

Deep-drawing

Drawing can also be used to pull metal over a die (male mold) to create a specific shape. For example, stainless steel kitchen sinks are formed by drawing the stainless steel sheet metal stock over a form (the die) in the shape of the sink. Beverage cans are formed by drawing aluminum stock over can-shaped dies. By comparison, hydroforming forces metal into a female mold using pressure.

There are many other manufacturing processes applied to sheet metal part design.

There are basically two approaches to sheet metal part design. One is referred to as the insert bends method; this is the traditional method. The other is the base flange method, which is more recent. The base flange approach is presented in this book because it is more natural when applied to sheet metal part design.

12.2 Sheet Metal Part Design Methodology: Base Flange

In this approach, the starting point is a base flange, which becomes the bottom face (a rectangle, 350 mm by 450 mm) of the cabinet being made.

1. Click the MY-TEMPLATE tab.
2. Double-click ANSI-MM-PART.
3. Click the front plane (or another appropriate plane).
4. Sketch and dimension the base flange, Sketch1 (a rectangle, 350 mm by 450 mm).
5. Click Base Flange from the Sheet Metal toolbar. The Base Flange PropertyManager appears (see Fig. 12-3).
6. Accept the default value for the sheet metal thickness. Notice that there is an option to use a gauge table.

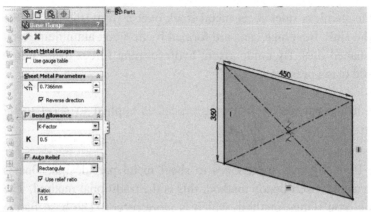

Figure 12-3 Sketch1, base flange

Base-Flange1 appears after Sheet-Metal1 in the FeatureManager (see Fig. 12-4).

Figure 12-4 FeatureManager

Create an edge flange

7. Click one of the edges of the base flange and then select the Line tool from the Sketch toolbar (see Fig. 12-5). A perpendicular plane for sketching the line is automatically selected; this is very useful in this approach.

8. Create an L-sketch, 100 mm long and 35 mm wide, Sketch2, with a fillet radius of 2.5 mm (see Fig. 12-6).

9. Exit sketch mode.

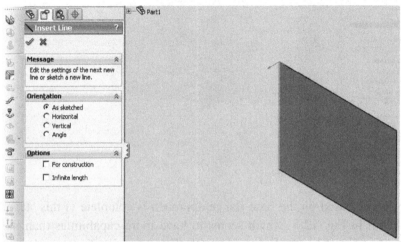

Figure 12-5 Effect of selecting the Line tool

Figure 12-6 Creating a sketch of the profile

Create miter flange features

10. Select Sketch2.

11. Click the Miter Flange button on the Sheet Metal toolbar, while the sketch is still active.

12. Select the four edges of the base flange, which are not automatically dimensioned, and set their height to 100 mm, as shown in Fig. 12-7.

13. Click OK from the Miter Flange PropertyManager.

14. Hide Plane1.

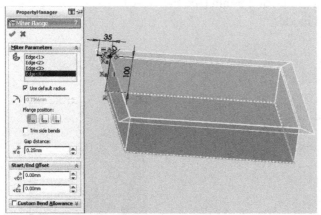

Figure 12-7 Miter flange feature

Our design based on the base flange approach is complete at this stage, as shown in Fig. 12-8, which seems to have more capabilities than the alternative insert bends design approach. Other venting features already discussed can now be included.

Figure 12-8 Completed part using base flange approach

12.3 Case Study Using Base Flange

A hanger support shown, in Fig. 12-9, is to be modeled. We will give a step-by-step procedure for the design. This hanger support will be required for finite element analysis to determine how it will function in real-life.

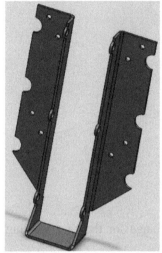

Figure 12-9 Sheet metal model of hanger support

1. Create a new part document.
2. Select the top view.
3. Sketch the base profile, Sketch1 (see Fig. 12-10).

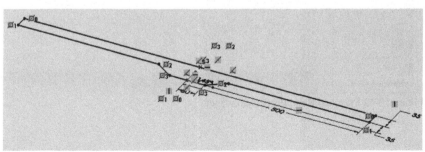

Figure 12-10 Sketch1

Base flange

4. With the base profile (Sketch1) active, click Base Flange/Tab (see Fig. 12-11).
5. Set the thickness to 5 mm.
6. Accept the K-Factor and Ratio default values.
7. Click OK to complete the base flange.

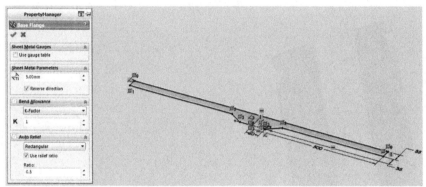

Figure 12-11 Base Flange PropertyManager

First sketched bending operation

8. Sketch a line, Sketch2, on the base flange for the first bending operation (see Fig. 12-12).
9. Click the fixed face (Face<1>) to the left of the line (Sketch2) in the Bend Parameters rollout.
10. Click OK to complete the sketched bend operation.

Figure 12-12 Sketched Bend PropertyManager

498

Second sketched bending operation
11. Sketch another line, Sketch2, on the base flange for the second bending operation (see Fig. 12-13).
12. Click the fixed face (Face<1>) to the right of the line (Sketch2) in the Bend Parameters rollout.
13. Click OK to complete the sketched bend operation.

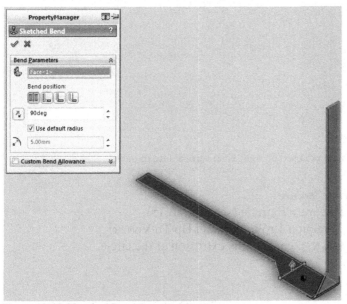

Figure 12-13 Sketched Bend PropertyManager

Create a circle on the right-hand vertical face
14. Click the right-hand vertical face of the model and use Normal To tool to position it appropriately (see Fig. 12-14).
15. Sketch a circle with a diameter of 25 mm on the right-hand vertical face.
16. Use the Smart Dimension tool to dimension the diameter as 25 mm, and the centre of the circle to be 20 mm from the angular vertex aligned with the edge.

Figure 12-14 Circle sketched on right-hand vertical face of model

Create an extrude feature

17. Click Feature > Extrude (see Fig. 12-15).
18. In the Direction 1 rollout, select Up To Vertex.
19. Click OK to complete this extrusion of the circle.

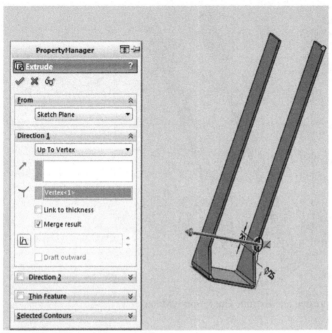

Figure 12-15 Extrude PropertyManager

Create another circle on the right-hand vertical face
20. Sketch a circle, concentric with the first, with a diameter of 15 mm (see Fig. 12-16).
21. Use the Smart Dimension tool to dimension the diameter as 15 mm.

Create an extruded cut
22. The second circle will now be cut out. Click Feature > Extrude Cut (see Fig. 12-16).
23. In the Direction 1 rollout, select Up To Vertex.
24. Click OK to complete this extrusion.

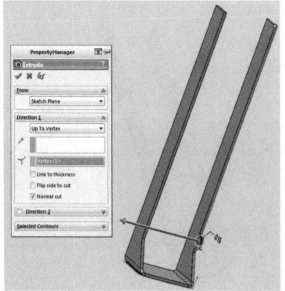

Figure 12-16 Circle created and Extrude Cut PropertyManager

Linear Pattern

25. The circle feature will be replicated. Click Features > Linear Pattern (see Fig. 12-17).
26. In the Direction 1 rollout, click the edge of the right-hand vertical side of the model.
27. Set the distance between patterns to 180 mm.
28. Set the number of instances to 3.
29. For the Features To Pattern, select Extrude3 and Extrude4.
30. Click OK to complete the pattern.

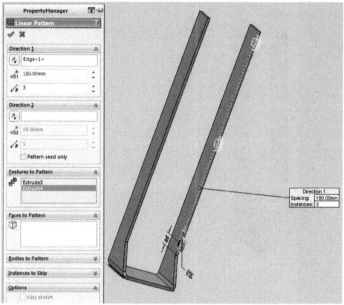

Figure 12-17 Linear Pattern PropertyManager

Mirror

31. The pattern of three circles will be mirrored. Click Features > Mirror (see Fig. 12-18).
32. For Mirror/Face Plane, select the Right Plane.
33. For Features to Mirror select LPattern2.
34. Click OK to complete the mirror operation.

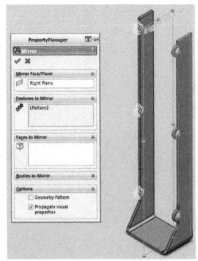

Figure 12-18 Mirror operation

Edge flange

35. Create a rectangular profile of length 390 mm from the top right corner (see Fig. 12-19).
36. Click the Edge-Flange tool (see Fig. 12-20).
37. For the Flange Parameter, click the edge of the rectangle common with the model.
38. Set the flange length to 75.00 mm.
39. Click OK.

Figure 12-19 Rectangular profile for edge flange

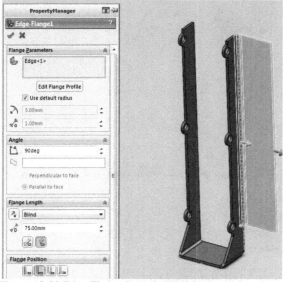

Figure 12-20 Edge Flange PropertyManager

Chamfer

40. Click Features > Chamfer (see Fig. 12-21).
41. Check Distance distance.
42. Select an edge and set the distance to 70.00 mm.
43. Click OK to complete the chamfer operation.

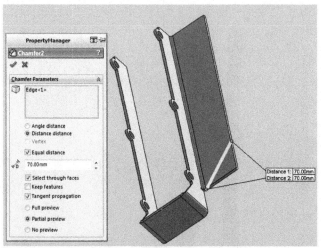

Figure 12-21 Chamfer PropertyManager

Create a circle on the front face of right-hand edge flange

44. Sketch a circle with a diameter of 33.4 mm, and centre on the chamfered edge (see Fig. 12-22).

45. Use the Smart Dimension tool to dimension the diameter as 33.4 mm and to set the centre a distance of 148.85 mm from the bottom of model.

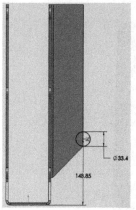

Figure 12-22 Circle on right-hand edge flange

Create an extruded cut

46. The circle will be cut out to form a notch. Click Feature > Extrude Cut (see Fig. 12-23).

47. In the Direction 1 rollout, select Up To Next.

48. Click OK to complete this extrusion.

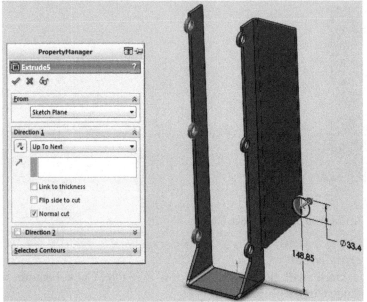

Figure 12-23 Extrude Cut PropertyManager

Linear Pattern

49. The semicircular notch will be replicated. Click Features > Linear Pattern (see Fig. 12-24).
50. In the Direction 1 rollout, click the edge of the right-hand vertical side of the model.
51. Set the distance between patterns to 130 mm.
52. Set the number of instances to 3.
53. For the Features To Pattern select Extrude5.
54. Click OK to complete the pattern.

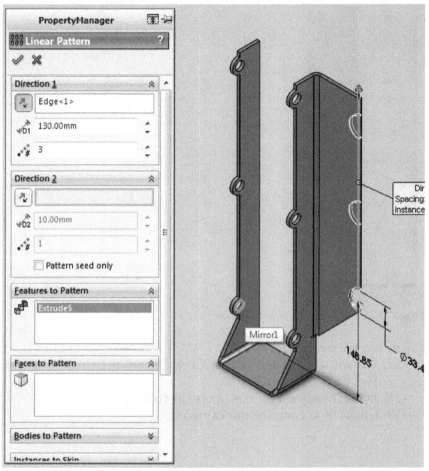

Figure 12-24 Linear Pattern PropertyManager

Create a circle on the front face of right-hand edge flange

55. Sketch two circles each of diameter 10 mm and with a vertical centre-centre distance of 35 mm (see Fig. 12-25).

56. Use the Smart Dimension tool for the dimensions shown.

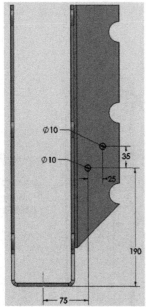
Figure 12-25 Two circles for an extruded cut

Create extrude cut feature

57. We will cut out the circles. Click Feature > Extrude Cut (see Fig. 12-26).
58. In the Direction 1 rollout, select Up To Next.
59. Click OK to complete this extrusion.

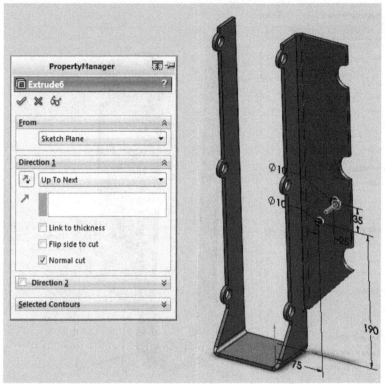

Figure 12-26 Extrude Cut PropertyManager

Linear pattern on right-hand edge flange

60. We will replicate the pair of circles. Click Features > Linear Pattern (see Fig. 12-27).

61. In the Direction 1 rollout, click the edge of the right-hand vertical side of the model.

62. Set the distance between patterns to 210 mm.

63. Set the number of instances to 2.

64. For the Features To Pattern select Extrude6.

65. Click OK to complete the pattern.

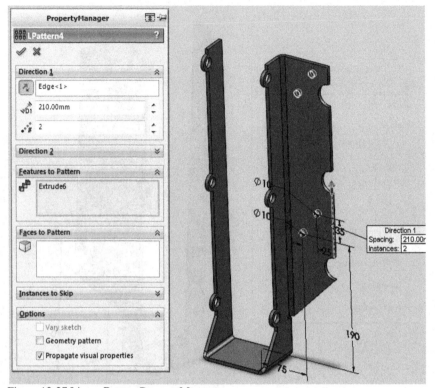

Figure 12-27 Linear Pattern PropertyManager

Mirror

66. We will mirror the notches and holes. Click Features > Mirror (see Fig. 12-28).
67. For the Mirror/Face Plane, select the Right Plane.
68. For the Features to Mirror select LPattern3, Extrude6, and LPattern4.
69. Click OK to complete the mirror operation.

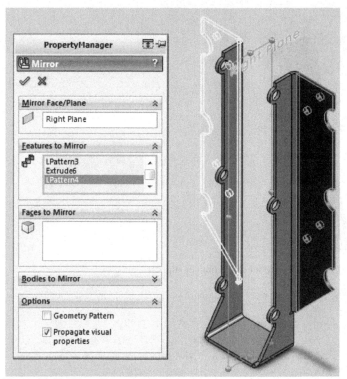

Figure 12-28 Mirror PropertyManager

The final sheet metal model of the hanger support is shown in Fig. 12-29. The list of features in FeatureManager are shown in Fig. 12-30.

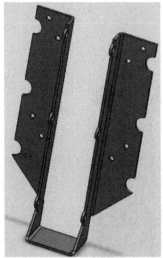

Figure 12-29 Final sheet metal model of the hanger support

Figure 12-30 FeatureManager for the sheet metal

Summary

We have used the base flange approach. It offers more flexibility because it does not require the part to be modeled first, which could be very challenging for complicated parts. It starts with a simple base, onto which other features, such as an edge flange or miter flange, can be added via a simple guide sketch. This approach forces the designer to *think sheet metal*, rather than to *think part design*

However, there are cases where the insert bends approach may be preferred, even though we have not discussed it. The two approaches can also be hybridized.

Chapter 13
Die Design

Objectives:

When you complete this chapter you will:

- Understand the scope of die design
- Understand the components of a die set
- Have designed a die holder of a die set
- Have designed a punch holder of a die set
- Have designed a die block
- Have designed a blanking punch
- Have designed a punch plate
- Have designed a pilot
- Have designed a stripper
- Have designed a back gage
- Have designed a front spacer
- Have designed a automatic stop
- Have assembled a design set

13.1 Scope of Die Design

Die design in a general sense involves either designing an entire press tool with all components taken together, or in a limited way it involves designing that particular component which is machined to receive the blank.

13.2 Components of a Die Set

Die sets are made by several manufacturers and they may be purchased in a great variety of shapes and sizes. The major components of a die set are the die holder (the lower part of the die) and the punch holder (the upper part of the die), as shown in Fig. 13-1. The *punch shank* **A** is clamped in the ram of the press, which is reciprocated up and down by a crank. In operation, the *punch holder* **B** moves up and down with the ram. *Bushings* **C**, pressed into the punch holder, slide on *guide posts* **D** to maintain the precise alignment of the cutting members of the die. The *die holder* **E** is clamped to the bolster plate (a thick steel plate fastened to the press frame) of the press by bolts passing through *slots* **F**.

The other components of a die set include the *die block, blanking punch, punch plate, pilot, stripper, back gage, front spacer,* and *automatic stop*. Each of these has to be designed and all components have to be assembled to cut a particular scrap strip.

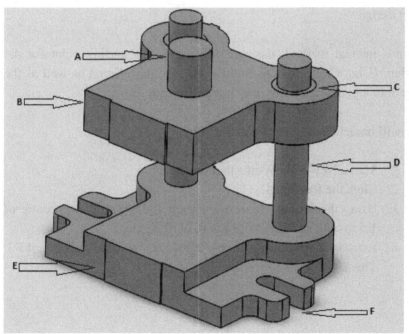

Figure 13-1 A typical die set

13.3   Pierce and Blank Die

The pierce and blank die is one of the most basic. The steps involved include the following:

1. Carefully study the part print since the information given on it provides many clues for solving the design problem.
2. Design a scrap strip as a guide for laying out the view of the actual dies.
3. Design the parts of the die set and assemble them.

In pierce and blank die operation, the die pierces two holes at the first station, and then the part is blanked out at the second station. The material from which the blanks are removed is cold-rolled steel strip.

Part design

With a manual method, the part drawing is the starting point for die design. Using CAD, such as SolidWorks, the part design as well as the part drawing, are fairly easy and straightforward.

We will model the part shown in Fig. 13-2.

1. Start a New SolidWorks Part document.
2. Click the front plane.
3. Using the Straight Slot tool, sketch a slot with centre-centre of 1.5 in with an arc radius of 9/16 in (0.563 in).
4. Extrude the basic profile by 0.0625 in, as shown in Fig. 13-3. This is the blanked feature.

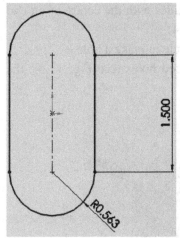

Figure 13-2 Basic profile for the part

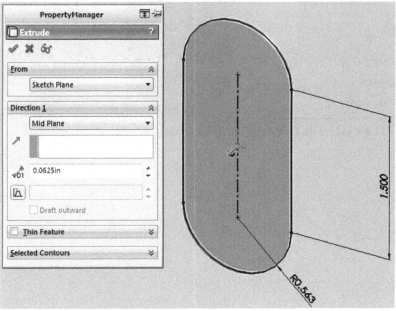

Figure 13-3 Basic profile extruded by 0.0625 in

5. Sketch a circle whose centre coincides with the centre of an end arc, 0.563 in diameter (see Fig. 13-4).

6. Use the Extrude Cut tool on the circle to make a hole.

7. Mirror the hole, to give two pierced holes (see Fig. 13-5); the blanked strip is shown in Fig. 13-6.

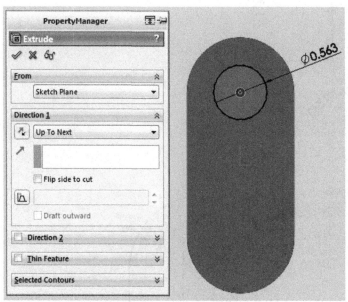

Figure 13-4 Extruded cut for the circle to define a hole

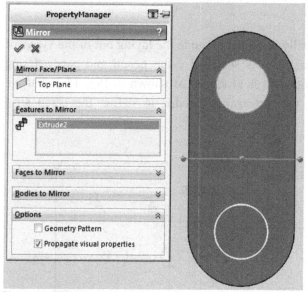

Figure 13-5 Mirrored hole

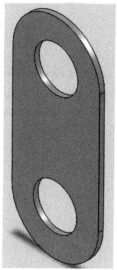

Figure 13-6 Pierced holes and blanked strip

13.4 Scrap Strip

A scrap strip is normally designed to guide the laying out of the views of the actual die. Manually, it is sketched using pencil and paper. Using SolidWorks, the scrap strip is produced as shown in Fig. 13-7, but we have only included the dimensions needed to define the scrap strip layout.

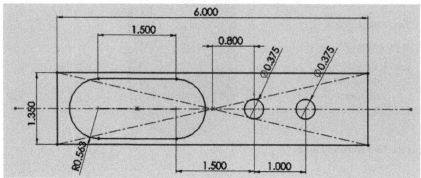

Figure 13-7 Scrap strip layout

13.5 Design of Die Holder of a Die Set

Designing the die holder of a die set is one of the major tasks in die design. Standard dimensions are available from manufacturers' websites.
1. Create the basic profile of a die holder on the top plane, as shown in Fig. 13-8.

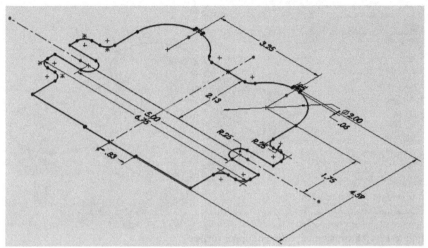

Figure 13-8 Sketch for die holder of a die set

2. Extrude the die holder through 0.502 in, as shown in Fig. 13-9.

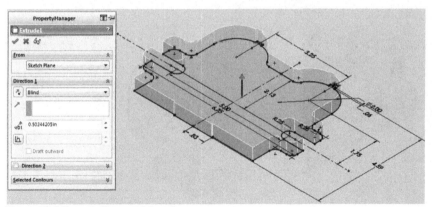

Figure 13-9 Extrude PropertyManager

3. Extrude an additional portion through 0.5 in, as shown in Fig. 13-10.

4. Create one hole of diameter 0.75 in.

5. Mirror the hole about the Right Plane as shown in Fig. 13-11. The final model is shown in Fig. 13-12.

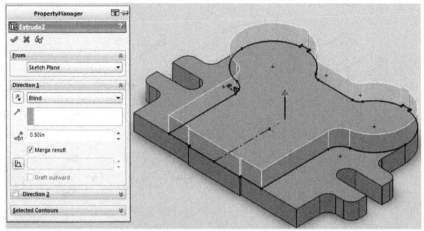

Figure 13-10 Extrusion of portion of upper surface

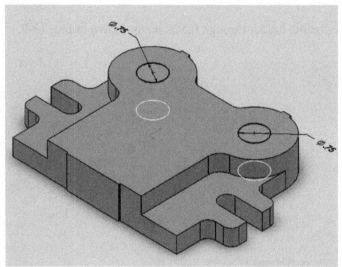

Figure 13-11 A hole created and mirrored about the right plane

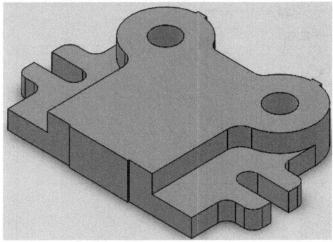

Figure 13-12 Die holder model

13.6   Design of Punch Holder of a Die Set

Designing the punch holder of a die set is another major task in die design. Standard dimensions are available from manufacturers' websites.

1. Create the basic profile of a die set punch holder on the top plane, as shown in Fig. 13-13.

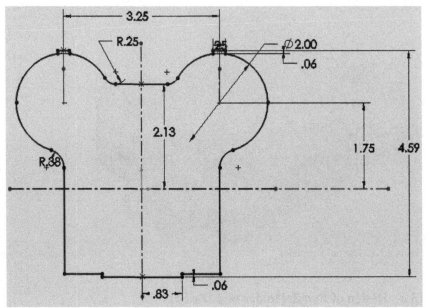

Figure 13-13 Basic profile of a die set punch holder

2. Extrude the basic punch holder sketch through 1.00 in, as shown in Fig. 13-14.

3. Create one hole, of diameter 1.00 in, as shown in Fig. 13-15.

4. Mirror the hole about the Right Plane.

5. Add the punch shank, as shown in Fig. 13-16.

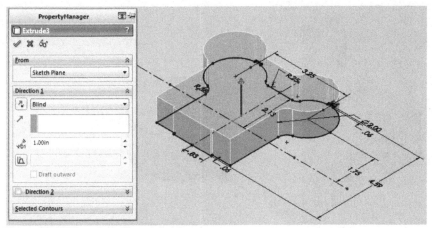

Figure 13-14 Extrusion of punch holder

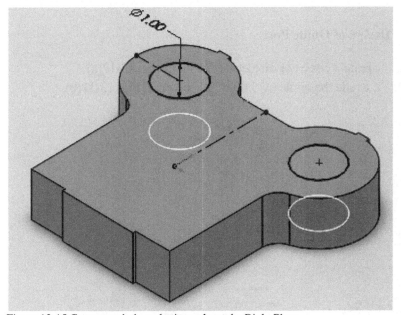

Figure 13-15 Create one hole and mirror about the Right Plane

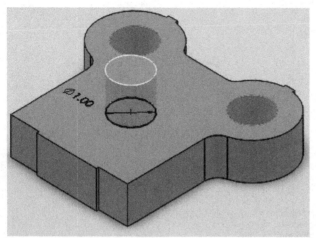

Figure 13-16 Add the punch shank

13.7 Design of Guide Post

1. Create a circle, of diameter 0.75 in, in Fig. 13-17(a).
2. Extrude the circle by 5.00 in, as shown in Fig. 13-17(b).

 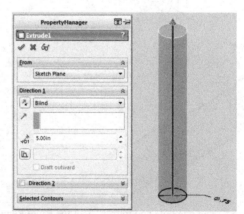

(a) Circle   (b) Extrusion
Figure 13-17 Guide post

13.8 Design of Bushing

1. Create two concentric circles, with diameters 0.75 in and 1.00 in, respectively, as shown in Fig. 13-18.
2. Extrude the circles through 1.00 in.

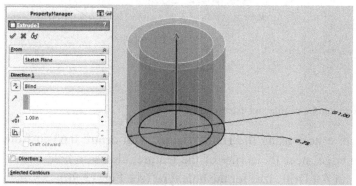

Figure 13-18 Extrude PropertyManager for bushing

13.9 Design of Die Block

1. Create a rectangle, 3.5 in by 3.0 in, as shown in Fig. 13-19.
2. Extrude the rectangle through 0.9375 in.

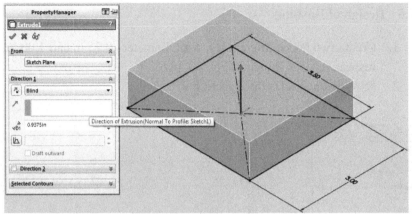

Figure 13-19 Rectangular profile and extrusion

3. Create a hole using the Hole Wizard, with centre 0.63 in from each edge, near the corner of the feature; choose a hole size of "I" and All Drill sizes for the hole type (see Fig. 13-20).

4. Linear Pattern/Mirror the hole to obtain the part. The final die block model is shown in Fig. 13-21.

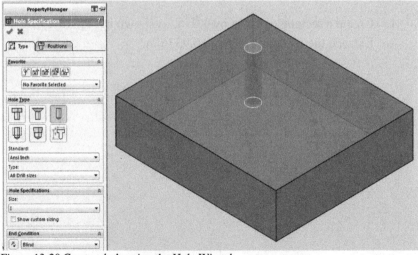

Figure 13-20 Create a hole using the Hole Wizard

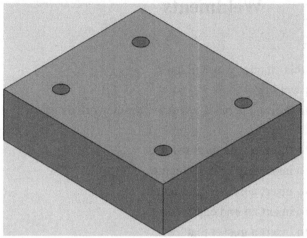

Figure 13-21 Die block model

Exercises

1. Design the blanking punch.
2. Design the punch plate.
3. Design the pilot.
4. Design the stripper.
5. Design the back gage.
6. Design the front spacer.
7. Design the automatic stop.
8. Assemble the components with the partial die set to complete the die set.
9. Produce drawings of the parts and an assembly drawing of the die set.

Chapter 14
Weldments

Objectives:

When you complete this chapter you will have:

- Learnt about weldment tools
- Learnt how to define the basic weldment framework using 3DSketch
- Learnt how to insert a structural member
- Learnt how to insert a fillet bead feature
- Learnt how to insert a trim/extend feature
- Learnt how to insert an end cap feature
- Learnt how to insert a gusset feature

14.1 Introduction

Weldments play a significant role in joining structural members, and, hence, are of vital importance in civil engineering. Also, weldments in manufacturing require specific attention. The functionality for weldments enables you to design a weldment structure as a single multi-body part. The basic approach is to use 2D and 3D sketches to define the basic framework. Then structural members are created using containing groups of sketch segments. Features that can be added using tools on the Weldments toolbar include gussets, end caps, etc.

14.2 Creating Parts with a 3D Sketch

Since the paths of weldments are mainly in 3D, a good grasp of 3DSketch tool is a prerequisite for creating weldments. From the main menu, select the 3DSketch tool (see Fig. 14-1) by clicking on the Sketch tab of CommandManager. Select the Line Tool. Alongside the cursor, the plane and the line will be displayed.

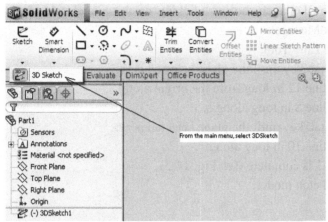

Figure 14-1 3DSketch tool

14.3 Tutorials

Tutorial 14.1: Bent tube

Model a bent tube, as three tubes, each with an outer diameter of 0.5 in and an inner diameter of 0.375 in, as shown in Fig. 14-2. Follow the route shown by the center lines in Fig. 14-2 and add 1 in radius fillets to the corners.

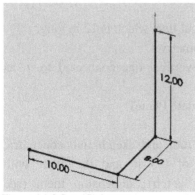

Figure 14-2 Specified route

1. Select the 3DSketch tool by clicking on the Sketch tab of the CommandManager.
2. Select the Line Tool.
3. Press the tab key until the plane selected is *yz*.
4. Sketch a line 12 in long from the origin along *y*.
5. Sketch a line 8 in long along *z*.
6. Press the tab key until the plane selected is *zx*.
7. Sketch a line 10 in long along *x*.
8. 3DSketch1 is complete (see Fig. 14-3).
9. Exit 3DSketch mode.

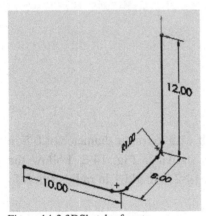

Figure 14-3 3DSketch of route

10. Click the top vertex of the vertical line, which is 12 in long.
11. Click Features > Reference Geometry.
12. Select the top vertex and the vertical line connected to it, as selections.
13. Select Normal to Curve to complete Plane1.
14. Select Plane1.
15. Right-click, select the Sketch tool and sketch two concentric circles, one with a diameter of 0.5 in and the other with a diameter of 0.375 in, as Sketch1; dimension them (see Fig. 14-4).

16. Exit sketch mode.

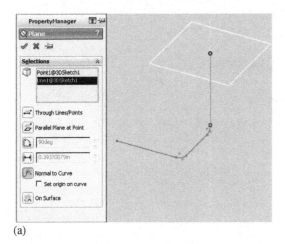

(a)

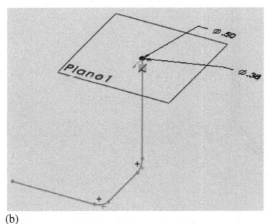

(b)

Figure 14-4 Normal plane and sketches for two circles

17. Click Features > Sweep.
18. Select Sketch1 as the Profile and 3DSketch1 as the Path (see Fig. 14-5).
19. Click OK.

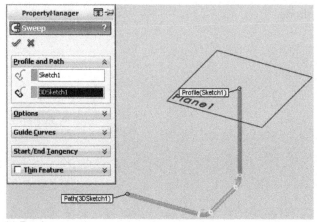

(a) Preview

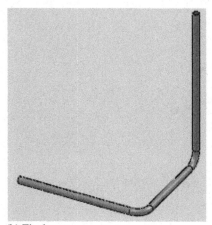

(b) Final part
Figure 14-5 Part for Tutorial 14.1

Tutorial 14.2: Bent tube

Model a bent tube, as three tubes, each with an outer diameter of 0.5 in and an inner diameter of 0.375 in, as shown in Fig. 14-6. Follow the route shown by the center lines in Fig. 14-6 and add 1 in radius fillets to the corners.

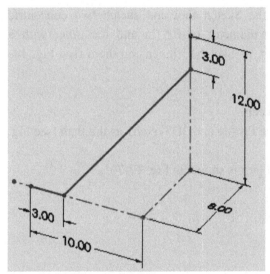

Figure 14-6 3DSketch of route

1. Select the 3DSketch tool by clicking on the Sketch tab of CommandManager.
2. Select the Line Tool.
3. Press the tab key until the plane selected is *yz*.
4. Sketch a line 3 in long from the origin along *y*.
5. Press the tab key until the plane selected is *zx*.
6. Sketch a line 3 in long along *x*.
7. Join the vertical and horizontal lines.
8. 3DSketch1 is complete (see Fig. 14-7(c)).

9. Exit 3DSketch mode.
10. Click the left vertex of the horizontal line which is 3 in long.
11. Click Features > Reference Geometry.
12. Select the top vertex and the vertical line connected to it, as selections.
13. Select Normal to Curve to complete Plane1.
14. Select Plane1.
15. Right-click, select the Sketch tool and sketch two concentric circles, one with a diameter of 0.5 in and the other with a diameter of 0.375 in, as Sketch1; dimension them (see Fig. 14-7(a)).
16. Exit sketch mode.
17. Click Features > Sweep.
18. Select Sketch1 as the Profile and 3DSketch1 as the Path (see Fig. 14-7(b)).
19. Click OK. The final part is shown in Fig. 17-7(d).

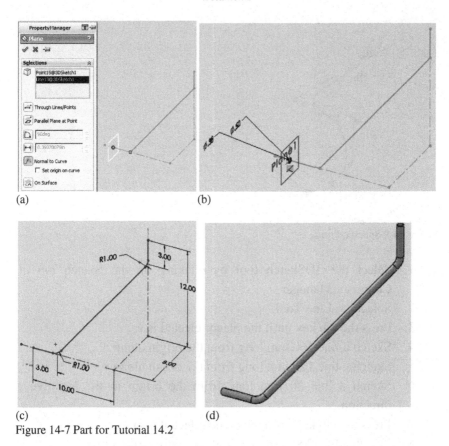

(a)   (b)

(c)   (d)

Figure 14-7 Part for Tutorial 14.2

Tutorial 14.3: Handlebar tube

Model the handlebar tube with 25.4 mm outer diameter and 23.4 mm inner diameter, as shown in Fig. 14-8. Follow the route shown by the center lines in Fig. 14-8 and add fillets to the corners as specified.

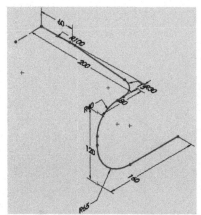

Figure 14-8 Specified route

1. Select the 3DSketch tool by clicking on the Sketch tab of CommandManager.
2. Select the Line Tool.
3. Press the tab key until the plane selected is *yz*.
4. Sketch a line 160 mm long from the origin along *z*.
5. Sketch a line 120 mm long from the origin along *y*.
6. Sketch a line 80 mm long from the origin in the negative *z* direction.
7. Press the tab key until the plane selected is *zx*.
8. Sketch a line 200 mm long in the negative *x* direction.
9. Add fillets to the corners as specified.
10. 3DSketch1 is complete (see Fig. 14-9).
11. Exit 3DSketch mode.

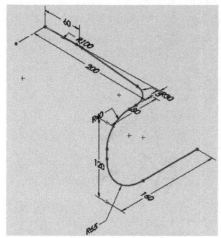

Figure 14-9 3DSketch of route

12. Click the end of the 160 mm long line.
13. Click Features > Reference Geometry.
14. Select the top vertex and the vertical line connected to it, as selections.
15. Select Normal to Curve to complete Plane1.
16. Select Plane1.
17. Right-click, select the Sketch tool and sketch two concentric circles, one with a diameter of 25.4 mm and the other with a diameter of 23.4 mm, as Sketch1; dimension them (see Fig. 14-10).
18. Exit sketch mode.

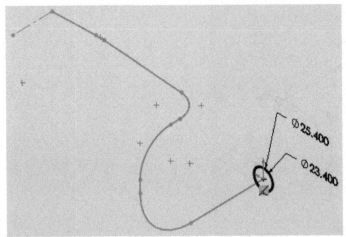

Figure 14-10 Normal plane and sketches for two circles

Figure 14-11 shows the preview while Fig. 14-12 shows the final model.

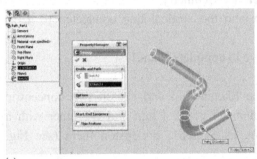

(a)

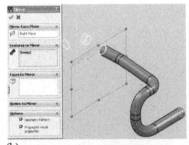

(b)

Figure 14-11 Preview of model

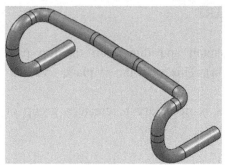

Figure 14-12 Final model of handlebars

14.4   Weldments Toolbar

Now that we have a good grasp of the 3DSketch tool, which is a prerequisite for creating weldments, we will study the Weldments toolbar.

The Weldments toolbar provides tools for creating weldment parts. The main tools are:

 Weldment

 Structural member

Gusset

 End cap

 Fillet bead

 Trim/extend

14.4.1 *Structural member*

The first step is to define the paths for which different profiles will be created. The paths are created using 3DSketch.

1.   Click the MY-TEMPLATE tab.

2. Double-click ANSI-MM-PART.
3. Click the top plane.
4. Using 3DSketch mode, sketch and dimension the top path (a rectangle, 20 in by 12 in). This is in the *xy*-plane (see Fig. 14-13).
5. Exit 3DSketch mode. This is necessary to create a group of structural members.

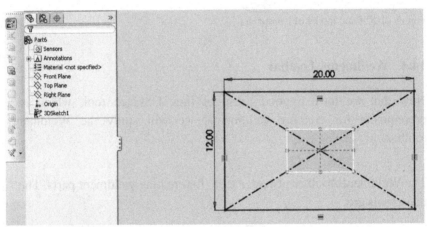

Figure 14-13 Top path (a rectangle, 20 in by 12 in) in the *xy*-plane

6. Click the top plane.
7. Click 3DSketch in CommandManager.
8. Pick one of the vertices of the top rectangular feature and sketch a line (switch to the *yz*-plane by pressing the tab key).
9. Sketch the other legs and use relations to make all four legs parallel and equal.
10. Dimension one of them to be 10 in.
11. Add relations so that one of the legs and the top members are perpendicular (90°) (see Figs 14-14 and 14-15). The complete sketch is shown in Fig. 14-15.

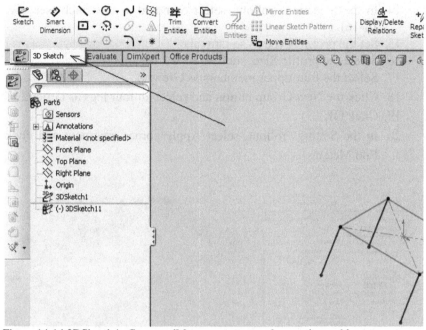

Figure 14-14 3DSketch in CommandManager necessary for creating weldments

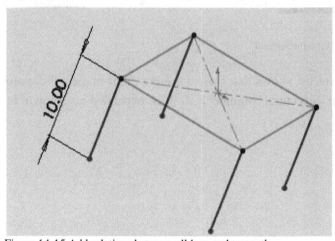

Figure 14-15 Add relations between all legs and top path

12. Exit 3DSketch mode.
13. Click the Structural Member toolbar.

14. In the Selections rollout, select Standard as ansi inch (see Fig. 14-16).
15. Set the profile Type as square tube.
16. Select the profile Size, such as 2 x 2 x 0.25.
17. Select the four upper members as Group1.
18. Click the New Group button and select all four legs as Group2.
19. Click OK.
20. In the Settings rollout, select Apply corner treatment and click End Meter.

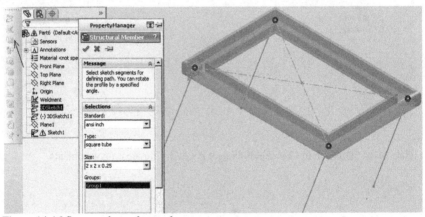

Figure 14-16 Structural member tool

Expanding the Cut list shows that there are eight multi-bodies; these are also included in Structural Member1 if it is expanded as shown in Fig. 14-17.

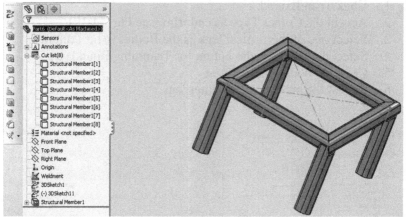

Figure 14-17 Cut list for eight multi-body structural members

14.4.2 *Trimming the structural members*

Trimming is needed because a solid body overlaps with or enters into another solid body, which is not allowed in practice. For example, if instead of defining Group1 and Group2 at the same time as defining the structural members, they are defined as two different structural members, then the need for trimming may arise. This scenario is shown in Fig. 14-18. The figure shows Structural Member1 and Structural Member2.

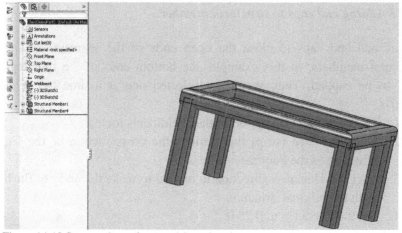

Figure 14-18 Structural member requiring trimming

1. Click Trim/Extend.
2. Accept the Corner Type as end trim (see Fig. 14-19).
3. Select one of the vertical legs as the Bodies to be Trimmed.
4. Select the Bodies radio button for Trimming Boundary.
5. Click the bounding surfaces.
6. Repeat the process for all other legs.

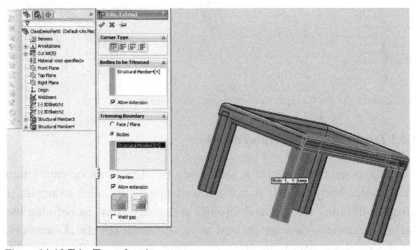

Figure 14-19 Trim/Extend option

14.4.3 *Adding end caps to structural members*

Let us add end caps to close the open ends of the segments of the structural members. In this example, the bottom ends of the four legs need to be capped. They should be selected one at a time (see Fig. 14-20).

1. Click the End Cap option of the Weldments toolbar.
2. Click the face (strip) that defines the cross section of the leg structure as the Parameter.
3. Sect the Thickness direction to inward to make the end cap flush with the original structure.
4. Set the thickness to 0.25 in.

5. Select Use thickness ratio.
6. Set the thickness ratio to 0.65 (try other values and observe the preview). The capped legs are shown in Fig. 14-21.

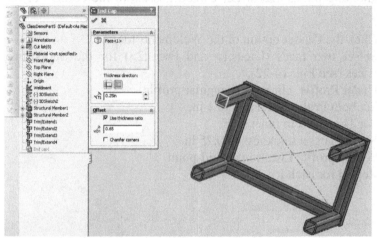

Figure 14-20 End cap option

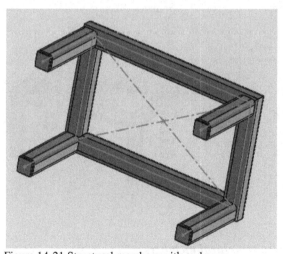

Figure 14-21 Structural members with end caps

14.4.4 *Adding gussets to structural members*

Gussets, which may have triangular or polygonal profiles, can be added to two adjoining planar faces. Let us add gussets to the corners of the four legs.

1. Click the Gusset option of the Weldments toolbar.
2. Select two faces (Face<1> and Face<2>) for the Supporting Faces (see Fig. 14-22).
3. Under Profile, click the triangular profile.
4. Set both Profile distances to 1 in.
5. Set Thickness to inner side.
6. Set the gusset thickness to 0.25 in.
7. Set the profile Location as mid-point.
8. Repeat for each leg.

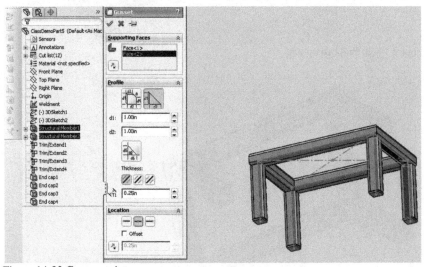

Figure 14-22 Gusset option

14.4.5 *Adding fillet beads to structural members*

Fillet beads can be added to two adjoining planar faces. Let us add fillet beads to the corners of the four legs.

1. Click the Fillet Bead option of the Weldments toolbar.

2. Click the face of the gusset.
3. Click two faces (Face<1> and Face<2>) for the Supporting Faces (see Fig. 14-23).
4. Repeat for each leg.

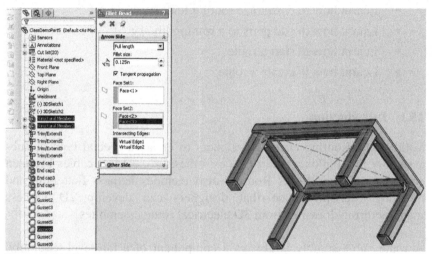

Figure 14-23 Fillet bead option

Chapter 15

Routings: Piping and Tubing

Objectives:

When you complete this chapter you will have:

- Learnt how to customize routing templates
- Learnt how to add parts to a routing library
- Learnt how to start a route
- Learnt how to create a route

15.1 Introduction

SolidWorks Routing enables designers to create a special type of sub-assembly that builds a path of pipes, tubes, or electrical cables between components. SolidWorks Routing also includes harness flattening and detailing capabilities, so that designers can develop 2D harness manufacturing drawings from 3D electrical route assemblies.

A route sub-assembly is always a component of a top-level assembly. When you insert certain components into an assembly, a route sub-assembly is created automatically. Unlike other types of sub-assemblies, you do not create a route assembly in its own window and then insert it as a component in the higher-level assembly.

Instead, you model the route by creating a 3D sketch of the centerline of the route path. SolidWorks generates the pipe, tube, or cable along the centerline.

SolidWorks makes extensive use of design tables to create and modify the configurations of route components. The configurations are distinguished by different dimensions and properties. If you are unfamiliar with these concepts, read Chapter 9.

15.2 Activating the SolidWorks Routing Add-In

The starting point for utilizing the routing tool is to activate the SolidWorks Routing add-in:
1. Click Tools > Add-Ins.

2. Select SolidWorks Routing to activate it for the current session in the Active Add-Ins.
3. Click OK to add SolidWorks Routing (see Fig. 15-1).

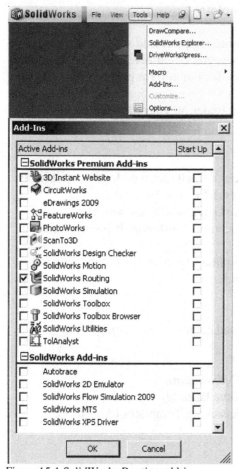

Figure 15-1 SolidWorks Routing add-in

15.3 Background

It should be noted that a route is always a component of a top-level assembly. This means that when we insert components into an assembly, a route sub-assembly is created automatically. There is a way of switching from the sub-assembly to the top-level assembly. We will consider this later.

15.4 Customizing Routing Templates

Now we will create a custom routing template and set its units to inches. If your organization's policy allows, you could save the custom template in the default template location, but for our current purpose, you can save it in a new folder that you create.

1. In Windows Explorer, create a folder on your local drive called *H:\MyRouting*.
2. In SolidWorks, click Open.
3. In the Open dialog:
    a. For Look in, browse to your default template location (typically *C:\Documents and Settings\All Users\Application Data\SolidWorks\SolidWorks<version>\templates*). If your default template location is different, browse to that location.
    b. In File of type, select Template (*.prtdot;*.asmdot;*.drwdot*).
    c. Select *routeAssembly.asmdot*.
    d. Click Open.

Now save a copy of the template and change some settings in it:
1. Click File, Save As.
2. In the Save As dialog:
    a. For Save in, browse to *H:\MyRouting*.
    b. For File name, type *MyRouteAssembly*.
    c. For Save as type, select Assembly Templates (*.asmdot*).
    d. Click Save.
3. Click Options.
4. In the dialog:
    a. On the Document Properties tab, select Units.
    b. Under Unit system, select IPS (inch, pound, second).
    c. Click OK.
5. Click Save (Standard toolbar).

15.5 Adding Parts to the Routing Library

The Routing Library contains parts (such as flanges, fittings, and pipes) for you to use in routes. By default, the Routing Library is located in a folder named *routing* in the Design Library. You can add components to existing folders in the Routing Library, or create new folders. You must have write access to your Design Library to create folders and add parts.

Create a new folder in the Routing Library and add an assembly fitting without acp assembly.
1. In the Task Pane:
    i.   Click the Design Library tab.
    ii. Browse to *Design Library\routing\assembly fittings*.
2. At the top of the task pane, click Create New Folder.
3. Type *MyLibrary* for the folder name and press Enter.

Let us consider some routing illustrations.

15.6 Illustration 1

Now add assembly fitting without *acp.sldasm* to the *MyLibrary* folder (see Fig. 15-2).
1. At the top of the task pane, click Add to Library.
2. In PropertyManager:
    i.   For Items to Add, select assembly fitting without *acp.sldasm* at the top of the flyout FeatureManager design tree.
    ii. Under Save To, make sure the *MyLibrary* folder is selected under the Design Library folder.
    iii. Click OK.
         The part is added to the Routing Library, and is available for selection when you create a route.
3. Close the part.

15.6.1 *Starting a route*

We will add some pipe and tube routes to an assembly.

1. Browse to *C:\Documents and Settings\All Users\Application Data\SolidWorks\SolidWorks<version>\design library\routing\assembly fittings\MyLibrary*.
2. Save the assembly as *MyAssyFitting.sldasm*.
   The assembly normally will already contain some fittings that need to be connected by pipe or tube routes.
3. Start the first route by dragging a flange into the assembly. Figure 15-2 shows a preview of the Design Library Assembly Fitting. You can use tools on the View toolbar to zoom, rotate, and pan the model view to facilitate working with the model.

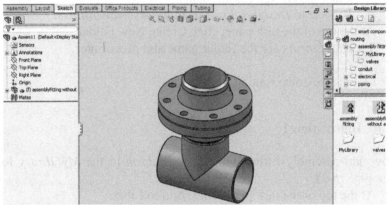

Figure 15-2 Assembly fitting without *acp.sldasm*

4. In the lower panel, double-click the *flanges* folder.
5. Click Piping > Start by Drag/Drop (see Fig. 15-3). The Design Library opens to the piping section of the Routing Library.
6. In the lower panel, double-click the *flanges* folder.
7. Drag slip on weld *flange.sldprt* from the library to the flange face on the regulator.
8. Drop the flange when it snaps into place (see Fig. 15-4).
9. In the dialog, select Slip On Flange 150-NPS2.
10. Click OK. The Route Properties PropertyManager appears.

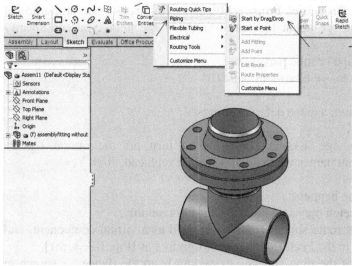

Figure 15-3 Drag/Drop option

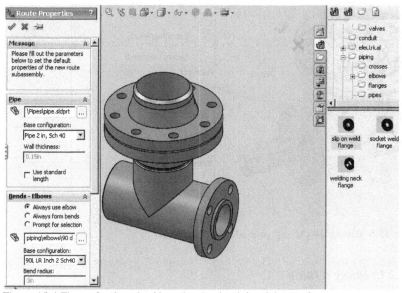

Figure 15-4 Flange for dragging/dropping on the right of the window

In the Route Properties PropertyManager, you specify the properties of the route you are about to create. Some of the items you can specify include:

- Which pipe or tube parts to use.
- Whether to use elbows or bends.

For this session, use the default settings.

11. Click OK.
12. If a message asks if you want to turn off the option Update component names when documents are replaced, click Yes.

The following happens:

- A 3D sketch opens in a new route sub-assembly.
- The new route sub-assembly is created as a virtual component, and appears in the FeatureManager design tree as [Pipe1-Assem1].
- A stub of the pipe appears, extending from the flange, as shown in Fig. 15-5.

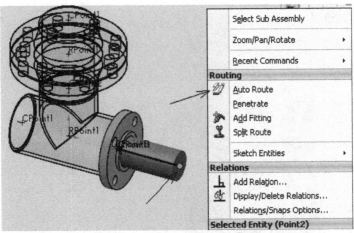

Figure 15-5 Starting a route: a pipe is added to the flange

15.6.2 *Creating a route*

We will start by creating the segments of the route.

1. Drag the endpoint of the stub to increase the pipe length, as shown in Fig. 15-6. You do not need to be exact.

2. If a message appears about not adding automatic relations, click OK. (The software is trying to add sketch relations to the weldment that is behind the pipe, but determines that the relations would over define the route sketch.)

Now we will add the horizontal flange to the route, so you can connect the pipe to it.

3. Zoom to the horizontal flange.
4. On the View menu, make sure Routing Points is selected and Hide All Types is cleared.
5. Move the pointer over the connection point (CPoint1) in the center of the flange. The pointer changes and the connection point is highlighted
6. Right-click CPoint1 and select Add to Route. A stub of the pipe extends from the flange.

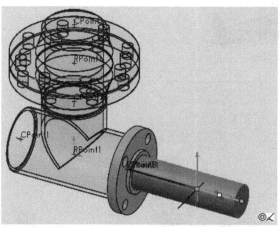

Figure 15-6 Pipe lengthens

15.7 Illustration 2

Let us switch to assembly *fitting.sldasm*. The sub-assembly has a tee and flanges (see Fig. 15-7). In this case, we can start our routing with three pipes. We will do the following:

1. Click CPoint1 and extend the pipe to the right (see Fig. 15-8).

2. Click CPoint2 and extend the pipe to the left.
3. Click CPoint1 on the right, and insert a line using 3DSketch along *y* on the *xy*-plane (see Fig. 15-9).
4. Click CPoint2 on the left, and insert a line using 3DSketch along *z* on the *yz*-plane.

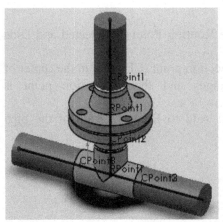

Figure 15-7 Assembly *fitting.sldasm*

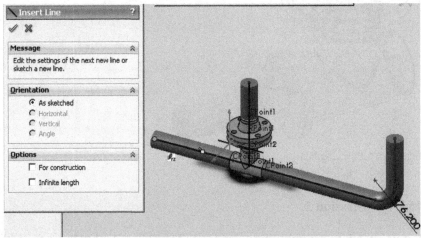

Figure 15-8 Extending the right-hand pipe vertically using the 3DSketch tool

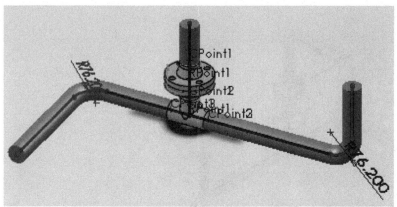

Figure 15-9 Extending the left-hand pipe horizontally using the 3DSketch tool

We will now add a pair of flanges to the vertical pipe and add an extra pipe above the pair of flanges.

5. Click the end (CPoint1) of the vertical pipe, and drag Slip On Flange to the point (CPoint1). Note that a pipe is automatically added to the pair of flanges (see Fig. 15-10).

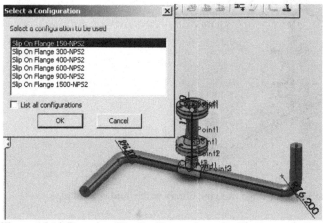

Figure 15-10 Dragging Slip On Flange from the library, to the top of the third vertical pipe

6. Click the end (CPoint1) of the vertical pipe, and extend it (see Fig. 15-11).
7. Click CPoint1, and insert a line using 3DSketch along *z* on the *yz*-plane.

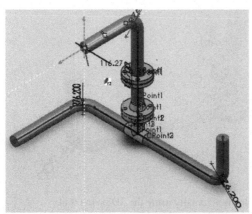

Figure 15-11 Extending the third pipe vertically along the *z*-direction on the *yz*-plane

8. Click the end (CPoint1) of the vertical pipe, and drag Slip On Flange to the point (CPoint1), as shown Fig. 15-12.

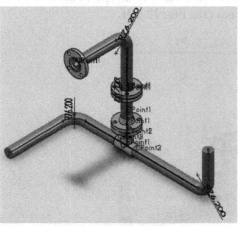

Figure 15-12 Dragging Slip On Flange from the library to the third vertical pipe

It should be noted that, so far, we have create a sub-assembly in a top-level assembly. The sub-assembly is shown in blue. To move to the top-level assembly, click Edit Assembly. This is shown in Fig. 15-13. The final model is shown in Fig. 15-14. The FeatureManager contains details of the routing operations.

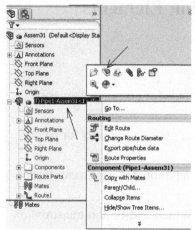

Figure 15-13 Switching to the top-level assembly

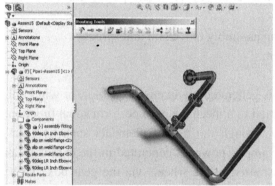

Figure 15-14 Final model

15.8   Route Drawing

When the routing exercise is complete, a new drawing document is opened to draw the route. The same procedure for drawing parts applies here, with the bill of materials added. This is left as an exercise.

Chapter 16

Power Transmission Elements

Objectives:

When you complete this chapter you will:

- Understand the available design tools for power transmission systems
- Understand how to size supporting plates and shafts for power transmission systems
- Be able to model an assembly of a pinion and gear transmission system
- Be able to model an assembly of a rack-and-pinion transmission system
- Be able to model an assembly of a belt-and-pulley transmission system
- Be able to model an assembly of a chain and sprocket transmission system

Power, that is movement, is generated by an engine, motor, or windmill. The power is then transmitted to a mechanism to perform some function. Transmission is generally through the use of elements such as gears, pulleys, and chains. For example, power is generated in the engine of an automobile, and then transmitted to the wheels via the gear box. These power transmission elements are considered in this chapter.

16.1 Gears and Power Transmission

A gear is a rotating machine part having cut teeth, or cogs, which mesh with another toothed part in order to transmit torque. Two or more gears working in tandem are called a transmission. They can produce a mechanical advantage through a gear ratio and thus may be considered a simple machine. Geared devices can change the speed, magnitude, and direction of a power source. The most common situation is for a gear to mesh with another gear; however, a gear can also mesh a non-rotating toothed part, called a rack, thereby producing translation instead of rotation.

Combining two gears with different numbers of teeth produces a mechanical advantage, where the rotational speeds and the torques of the two gears differ through a simple relationship.

In transmissions that offer multiple gear ratios, such as bicycles and cars, the term gear, as in first gear, refers to a gear ratio rather than an actual physical gear. The term is used to describe similar devices even when the gear ratio is continuous rather than discrete, or when the device does not actually contain any gears, as in a continuously variable transmission.

The gears in a transmission are analogous to the wheels in a pulley. The advantage of gears is that the teeth prevent slippage.

The remaining sections of this chapter discuss the methodology for the assembly modeling of a pinion and gear, rack and pinion, belt and pulley, and chain and sprocket transmission systems.

16.2   Spur Gears

Spur gears or straight-cut gears are the simplest type of gear. They consist of a cylinder or disk and the teeth project radially. Although they are not straight-sided in form, the edge of each tooth is straight and aligned parallel to the axis of rotation. These gears can be meshed together correctly only if they are fitted on parallel axles.

16.2.1 *Creating gears*

SolidWorks can be used to create different types of gears such as spur, bevel, and helical gears. As many as eighteen standards are available in the SolidWorks Toolbox (see Fig. 16-1). To access these standards, click Design Library > Toolbox.

The Toolbox offers quite a number of design tools for bearings, bolts and screws, jig bushings, keys, nuts, O-rings, power transmission elements (chain sprockets, gears, timing belt pulleys), retaining rings, structural members, and washers, as shown in Fig. 16-2. These design tools help the design engineer to work more effectively and efficiently. In this chapter, we discuss the use of SolidWorks for modeling power transmission elements.

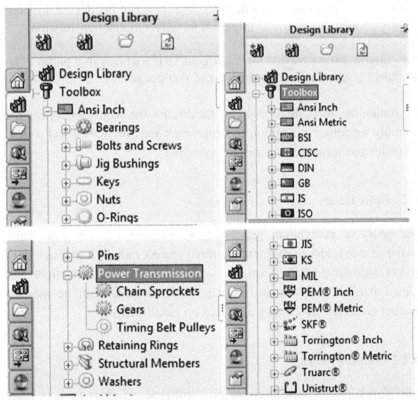

Figure 16-1 SolidWorks Toolbox standards   Figure 16-2 Toolbox design tools

16.2.2 *Power description*

We will model a gear system with a pinion (Gear 1) and gear (Gear 2). Their specifications are:

> *Gear 1: pinion*
> Diametral pitch = 24
> Number of teeth = 30
> Face thickness = 0.5 in
> Bore diameter = 0.5 in
> Hub diameter = 1.0 in
> Pressure angle = 20°
>
> *Gear 2: gear*
> Diametral pitch = 24
> Number of teeth = 60
> Face thickness = 0.5 in
> Bore diameter = 0.5 in
> Hub diameter = 1.0 in
> Pressure angle = 20°

Using SolidWorks, we will create an assembly model consisting of the two spur gears, two pins and a support plate, and animate the movement of the gears.

16.2.3 *Support plate sizing*

The pitch diameters for the pinion and gear are respectively:

$$D_p = 30/24 = 1.25''$$
$$D_g = 60/24 = 2.50''$$

The support plate dimensions are given by:

$$W = (D_g + 2 \times 0.5)$$
$$L = \left[(D_p + 0.5) + (D_g + D_p)/2 + W/2\right]$$

Leading to W = 3.5 in and L = 5.375 in.

16.2.4 *Gear assembly modeling*

There are three main steps in creating the gear assembly model: modeling of the support plate, modeling of the pin, and assembly of the support plate, pin, and gears.

Support plate model
1. Create a new part document.
2. Select the top plane.
3. Sketch a rectangular profile, 5.4 in by 3.5 in, Sketch1 (see Fig. 16-3).
4. Extrude Sketch1 through 0.5 in to get Extrude1.
5. Create a circle, Sketch2, on top of Extrude1 with centre (1.75, 1.00), as shown in Fig. 16-4.
6. Extrude-cut this circle through the thickness, to get Extrude2.

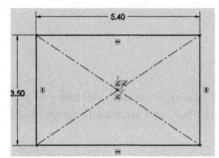

Figure 16-3 Sketch1

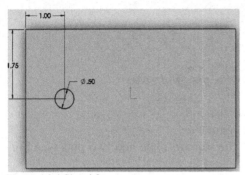

Figure 16-4 Sketch2

Create a linear pattern of Extrude2

7. Click Features > Linear Pattern. The Linear Pattern PropertyManager is displayed, as shown in Fig. 16-5.
8. Click a horizontal edge (Edge <1>) as Direction 1.
9. Set the number of instances to 2.
10. Select Extrude2 as the Features to Pattern.
11. Click OK.
12. Save the document.

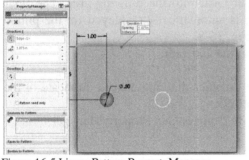

Figure 16-5 Linear Pattern PropertyManager

Pin model

1. Create a new part document.
2. Select the top plane.
3. Sketch a circular profile of diameter 0.5 in, Sketch1.
4. Extrude Sketch1 through 1.5 in to get Extrude1.

5. Click OK.
6. Save the document.

16.2.5 *Assembly of the support plate, pin, and gears*

Assembly of the support plate and two pins
1. Create a new assembly document.
2. Create an assembly with the support plate and two pins (see Fig. 16-6).

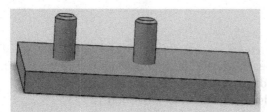

Figure 16-6 Assembly of the support plate and two pins

Adding gears to the assembly of the support plate and two pins
3. Click Design Library > Toolbox > Ansi Inch > Power Transmission > Gears (see Fig. 16-7).
4. Click and drag the Spur Gear into the drawing area. The Spur Gear PropertyManager is automatically displayed for the pinion and gear, as shown in Fig. 16-8.
5. Enter the data for the pinion and gear.
6. Click OK.
7. Click any point on the graphics drawing area (a duplicate of the gear appears; see Fig. 16-9).
8. Click Cancel in the Insert Components PropertyManager.

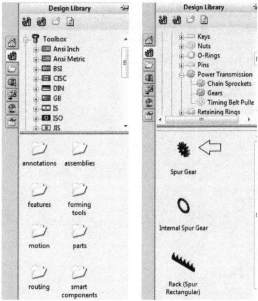

Figure 16-7 Toolbox for gears in the SolidWorks Design Library

Figure 16-8 Spur Gear PropertyManager showing pinion and gear data

Figure 16-9 Spur gears

Ordinary mating
9. Rotate the pinion and gear.
10. Click Mates.
11. Select the hole in the pinion and the outer surface of a pin and apply a concentric mate.
12. Select the top of the pinion and the top surface of the support and apply a distance mate of 1.0 in (see Fig. 16-10).

13. Repeat for the gear (see Fig. 16-11). Figure 16-12 shows the meshed pinion and gear.

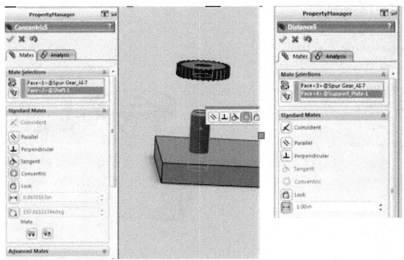

Figure 16-10 Mating conditions for pinion

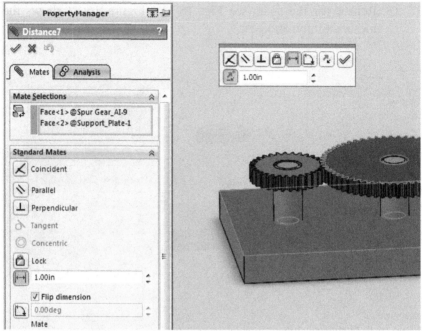

Figure 16-11 Mating conditions for gear

Figure 16-12 Meshed pinion and gear

Mechanical mating

14. Click Mates > Mechanical Mates (see Fig. 16-13 for the GearMate PropertyManager).

15. Click Gear.

16. Select the inner surface of the pinion hole and the inner surface of the gear hole.
17. Enter the ratios as 0.5 and 1.
18. Click OK.

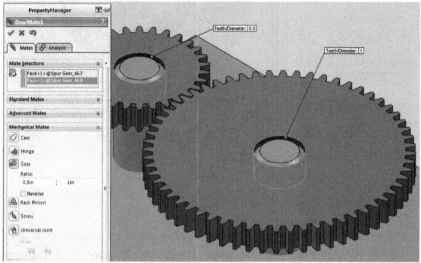

Figure 16-13 Mechanical mating conditions

16.2.6 *Animation*

1. Click the Motion Study1 icon at the bottom right corner of the graphics window.
2. Click the motor icon (see Fig. 16-14).
3. Click the top of the pinion as the component.
4. Set the motion speed as required (e.g. 3000 rpm).

Figure 16-14 Animation of pinion and gear

16.3 Rack-and-Pinion Gears

A rack is a toothed bar or rod that can be thought of as a sector gear with an infinitely large radius of curvature. Torque can be converted to linear force by meshing a rack with a pinion: the pinion turns and the rack moves in a straight line. Such a mechanism is used in automobiles to convert the rotation of the steering wheel into the left-to-right motion of the tie rod(s). Racks also feature in the theory of gear geometry, where, for instance, the tooth shape of an interchangeable set of gears may be specified for the rack (infinite radius), and the tooth shapes for gears of particular actual radii then derived from that. A type of rack-and-pinion gear is employed in a rack railway.

16.3.1 *Problem description*

A rack-and-pinion system consists of the rack and pinion and their specifications are:

> *Rack*
> Diametral pitch = 24
> Face width = 0.25 in
> Pitch height = 1.5 in
> Length = 5 in
> Pressure angle = 14.5°

> *Pinion*
> Diametral pitch = 24
> Number of teeth = 20
> Face width = 0.25 in
> Hub style: One side
> Hub diameter = 1.0 in
> Overall length = 1.0 in
> Nominal shaft diameter = 1/2 in
> Pressure angle = 14.5°

> *Shaft*
> Shaft diameter = 1/2 in
> Shaft length = 2.25 in

Using SolidWorks, we will create an assembly model consisting of the rack, pinion and shaft, and animate the movement of the gears.

16.3.2 *Gear assembly modeling*

There four main steps in creating the gear assembly model: insert the shaft, insert the pinion from the Design Library and assemble it to the shaft to form a sub-assembly, insert the rack from the Design Library, and the final assembly of the sub-assembly and rack.

Insert shaft into assembly
1. Create a new assembly document.
2. Click Assembly > Insert Components.
3. Insert the shaft (see Fig. 16-15).

Figure 16-15 Shaft inserted first in the assembly

Sub-assembly of shaft and pinion from the Design Library
4. Click Design Library > Toolbox > Ansi Inch > Power Transmission > Gears (see Fig. 16-16).
5. Click and drag the Spur Gear into the drawing area. The Spur Gear PropertyManager is automatically displayed for the pinion and gear, as shown in Fig. 16-17.
6. Enter the data for the spur gear (pinion).
7. Click OK.

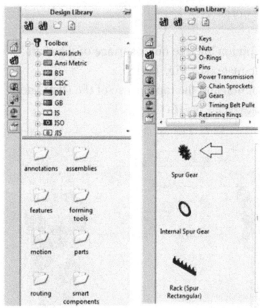

Figure 16-16 Toolbox for gears in the SolidWorks Design Library

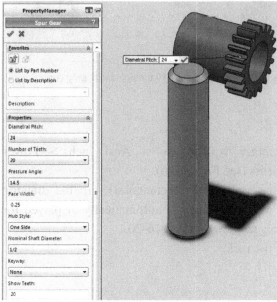

Figure 16-17 Spur Gear PropertyManager

8. Rotate the pinion appropriately.
9. Click Mates.
10. Select the hole in the pinion and the outer surface of the shaft and apply a concentric mate (see Fig. 16-18).
11. Select the top of the pinion and the top surface of the shaft and apply a coincident mate (see Fig. 16-19).

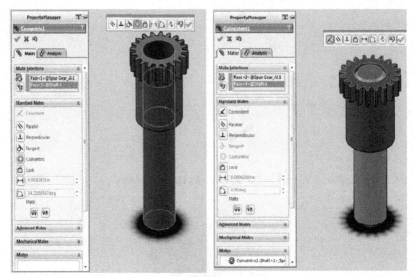

Figure 16-18 Concentric mating condition   Figure 16-19 Coincident mating condition

Insert rack into assembly

12. Click Design Library > Toolbox > Ansi Inch > Power Transmission > Gears (see Fig. 16-16).
13. Click and drag the rack (Spur Rectangular) into the drawing area. The Rack PropertyManager is automatically displayed for the pinion and gear, as shown in Fig. 16-20.
14. Enter the data for the rack.
15. Click OK.

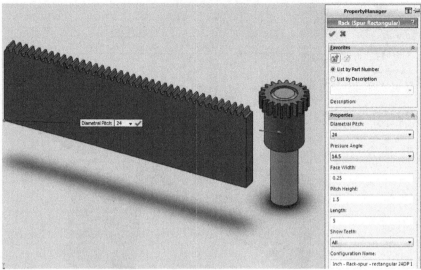

Figure 16-20 Rack PropertyManager

Ordinary mating

16. Click Mates.
17. Select the top of the pinion and the outer flat surface of the rack and apply a parallel mate (see Fig. 16-21).
18. Select the top of the pinion and the outer flat surface of the rack and apply a coincident mate (see Fig. 16-22).
19. Select the bottom tooth of the rack and the top tooth of the pinion and apply a tangent mate (see Fig. 16-23).

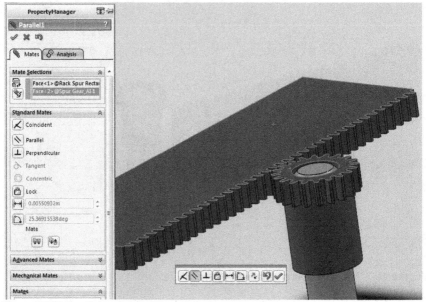

Figure 16-21 Parallel mate condition

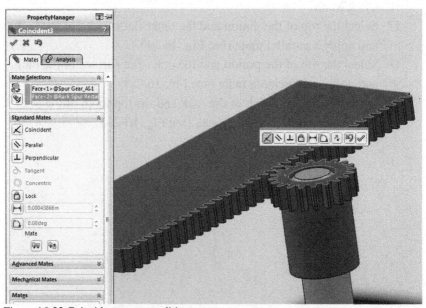

Figure 16-22 Coincident mate condition

Figure 16-23 Tangent mate condition

Mechanical mating

20. Click Mates > Mechanical Mates (see Fig. 16-24 for the Rack Pinion PropertyManager).
21. Select Rack Pinion.
22. Select the top edge of the rack and the side face of the pinion.
23. Click OK (see Fig. 16-25 for the final assembly).
24. Save the assembly.

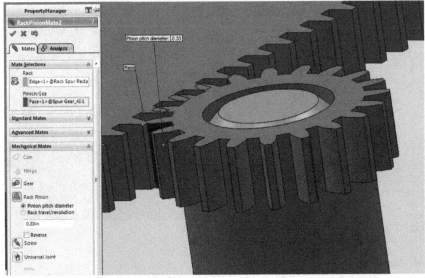

Figure 16-24 Mechanical mate condition

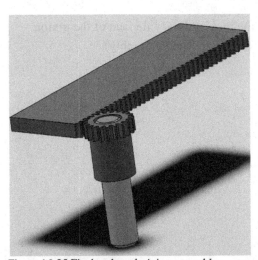

Figure 16-25 Final rack-and-pinion assembly

The FeatureManager is shown in Fig. 16-26.

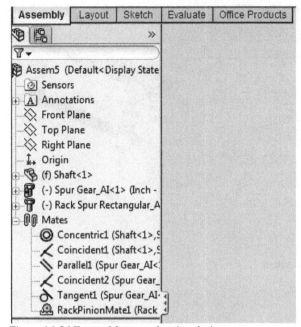

Figure 16-26 FeatureManager showing design tree

16.3.3 *Animation*

25. Click the Motion Study1 icon at the bottom right corner of the graphics window.
26. Click the motor icon (see Fig. 16-27).
27. Click the pinion as the component.
28. Set the motion speed as required (e.g. 100 rpm).
29. Click OK (see Fig. 16-28 for the position of the pinion at the end of the simulation).

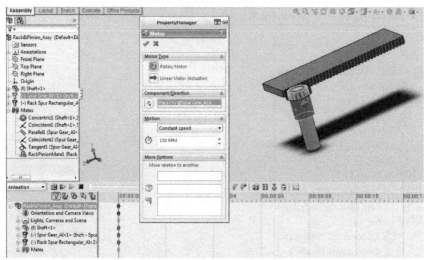

Figure 16-27 Motion of gear system

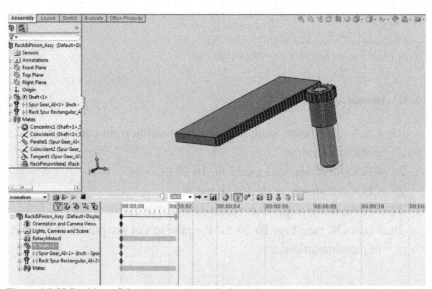

Figure 16-28 Position of the pinion at the end of simulation

16.4 Belts and Pulleys

A belt is the cheapest way to transmit power between two or more shafts that are not axially aligned. Power transmission is achieved by specially designed belts and pulleys. The demands on a belt-drive transmission system are large and this has led to many variations on the theme. They run smoothly and with little noise, and cushion motor and bearings against load changes, albeit with less strength than gears or chains. However, improvements in belt engineering allow use of belts in systems that only formerly allowed chains or gears.

A belt is a loop of flexible material used to link two or more rotating shafts mechanically. Belts may be used as a source of motion to transmit power efficiently or to track relative movement. Belts are looped over pulleys. In a two pulley system, the belt can either drive the pulleys in the same direction, or the belt may be crossed so that the direction of rotation of the shafts is opposite. A conveyor belt is one application where the belt is adapted to continually carry a load between two points.

16.4.1 *Problem description*

A belt-and-pulley system consists of a support plate and shaft, pulley, and belt. The specifications are:

Support plate
Length = 6.00 in
Width = 4.00 in
Thickness = 0.5 in
Hole diameter = 0.375 in. Two holes positioned diametrically, one is at (1.00, 2.00), the other is located 3.00 in away to the right (see Fig. 16-29).

Shaft
Length = 1.75 in
Diameter = 0.375 in

Pulley
Belt pitch = (0.200) - XL
Belt width = 0.38 in
Pulley style: Flanged
Number of grooves = 20
Hub diameter = 0.375 in
Overall length = 0.5 in
Nominal shaft diameter = 3/8 in
Keyway: None

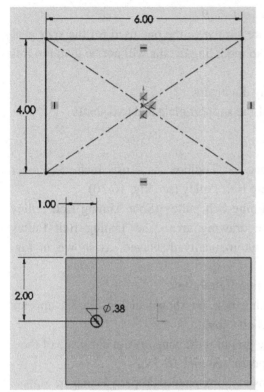

Figure 16-29 Support plate dimensions

Using SolidWorks, we will create an assembly model consisting of the support plate and shaft, pulley, and belt, and animate the movement of the belt and pulley.

16.4.2 *Belt-and-pulley assembly modeling*

There are three main steps in creating the belt-and-pulley assembly model: inserting the sub-assembly of the support plate and shaft, inserting the pulley, and inserting the belt.

Sub-assembly of support plate and shaft

The support plane and shaft are similar to the one used for the spur gear except for the specifications, so modeling details will not be given in this section.

1. Create a new assembly document.
2. Create an assembly of the support plate and two shafts.

Inserting the timing belt pulley

3. Click Design Library > Toolbox > Ansi Inch > Power Transmission > Timing Belt Pulley (see Fig. 16-30).
4. Click and drag the timing belt pulley (Spur Timing Belt Pulley Rectangular) into the drawing area. The Timing Belt Pulley PropertyManager is automatically displayed, as shown in Fig. 16-31.
5. Enter the data for Timing Belt Pulley.
6. Click OK. If a warning box, as shown in Fig. 16-32, appears accept the error and click Close.
7. Click any point on the graphics drawing area (a duplicate of the timing belt pulley appears; see Fig. 16-33).
8. Click Cancel in the Insert Components PropertyManager.

Figure 16-30 Toolbox for timing belt pulley in the SolidWorks Design Library

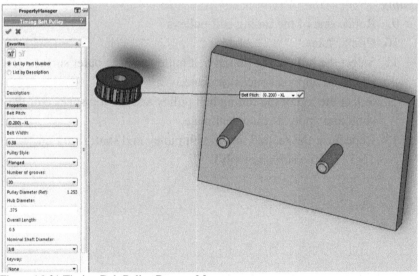

Figure 16-31 Timing Belt Pulley PropertyManager

Figure 16-32 Warning box

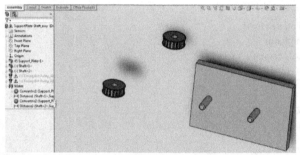

Figure 16-33 Timing belt pulleys, support plate, and shafts

Mating

9. Rotate one of the timing belt pulleys.

10. Click Mates.

11. Select the hole in the timing belt pulley and the outer surface of a shaft and apply a concentric mate.

12. Select the top of the timing belt pulley and the top surface of the shaft and apply a coincident mate.

13. Repeat for the second timing belt pulley and shaft (see Fig. 16-34 for the sub-assembly).

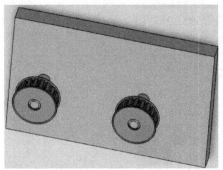

Figure 16-34 Sub-assembly of support plate, shafts, and timing belt pulleys

Inserting the belt

14. Click Insert > Assembly Feature > Belt/Chain (see Fig. 16-35 for the Belt/Chain insertion tool).

15. Click the top surface of a tooth on the left pulley and the top surface of a tooth on the right pulley to define the Belt Members (see Fig. 16-36).

16. In the Properties rollout, check Use belt thickness and enter 0.14 in.

17. Check Engage belt and Create belt part. A belt is created, as shown in Fig. 16-37.

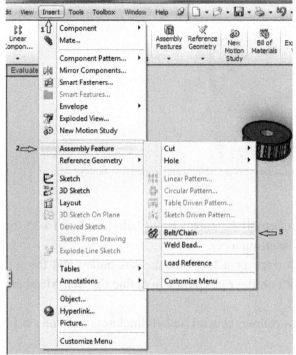

Figure 16-35 Belt/Chain insertion tool

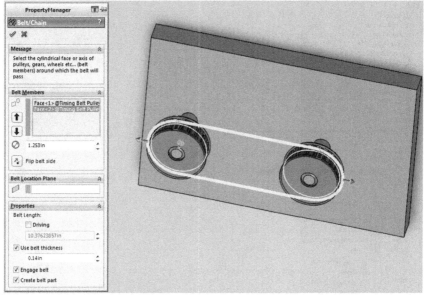

Figure 16-36 Belt Members defined

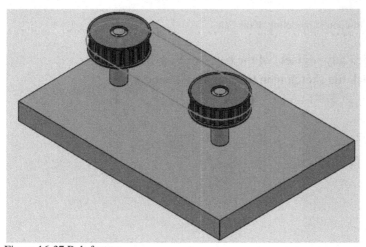

Figure 16-37 Belt feature

18. Expand Belt1 in the FeatureManager.
19. Right-click [Belt1^...]<1>->? and click the edit part icon (see Fig. 16-38).

Figure 16-38 In-context modeling of the belt

20. Click any segment of the belt (see Fig. 16-39).
21. Click the sketch icon to enter sketch mode.

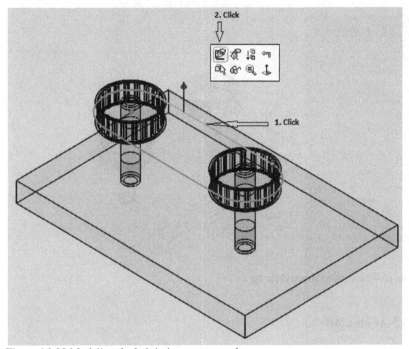

Figure 16-39 Modeling the belt in in-context mode

22. Click Features > Extrude Boss/Base. The Extrude PropertyManager is displayed, as shown in Fig. 16-40.
23. In the Direction 1 rollout, select Mid Plane.
24. Set the extrusion distance to 0.42 in.
25. Check the box for Thin Feature and set the thickness to 0.14 in.
26. Click OK.
27. Click Edit Component to exit editing the assembly.

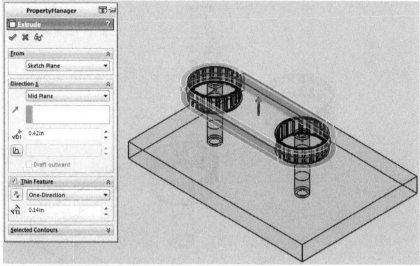

Figure 16-40 Extrude PropertyManager

16.4.3 *Animation*

1. Click the Motion Study1 icon at the bottom right corner of the graphics window.
2. Click the motor icon (see Fig. 16-41).
3. Click one of the pulleys as the component.
4. Set the motion speed as required (e.g. 100 rpm).
5. Click OK.

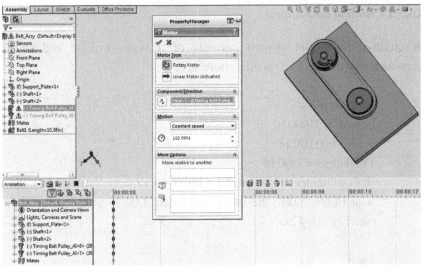

Figure 16-41 Motion PropertyManager

16.5   Chain Drive: Chains and Sprockets

A chain drive is a way of transmitting mechanical power from one place to another. It is often used to convey power to the wheels of a vehicle, particularly bicycles and motorcycles. It is also used in a wide variety of machines besides vehicles. Most often, the power is conveyed by a roller chain, known as the drive chain or transmission chain, passing over a sprocket gear, with the teeth of the gear meshing with the holes in the links of the chain. The gear is turned, and this pulls the chain putting mechanical force into the system.

SolidWorks creates chains and sprockets in a manner similar to that used for creating belts and pulleys.

16.5.1 *Problem description*

A chain-and-sprocket system consists of a support plate and shaft, sprocket, and chain. The specifications are:

Support plate
Length = 20.00 in
Width = 10.00 in
Thickness = 1.00 in
Hole diameter = 1.00 in. Two holes positioned diametrically, one is at (4.00, 5.00), the other is located 3.00 in away to the right (see Fig. 16-42).

Shaft
Length = 4.00 in
Diameter = 1.00 in

Sprocket
Chain number = SC610
Number of teeth = 24
Belt width = 0.38 in
Hub style: None
Nominal shaft diameter = 1 in
Keyway: None

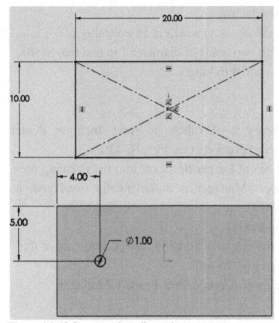

Figure 16-42 Support plate dimensions

Using SolidWorks, we will create an assembly model consisting of the support plate and shaft, sprocket, and chain, and animate the movement of the chain and sprocket.

16.5.2 Chain-and-sprocket assembly modeling

There are three main steps in creating the chain-and-sprocket assembly mod: inserting the sub-assembly of the support plate and shaft, inserting the sprocket, and inserting the chain.

Sub-assembly of support plate and shaft
The support plane and shaft are similar to the ones used for the spur gear except for the specifications, so the modeling details will not be given in this section.

1. Start a new assembly document.
2. Open *SupportPlate-Shaft_assy*, which is an assembly of a 20x10x1 in plate with two holes of diameter 1 in and two shafts, each of diameter 1in and 4 in long.

Inserting the sprocket
3. Click Design Library > Toolbox > Ansi Inch > Power Transmission > Chain Sprockets (see Fig. 16-43(a)).
4. Click and drag the Silent Larger Sprocket into the drawing area. The Sprocket PropertyManager is automatically displayed, as shown in Fig. 16-43(b).
5. Enter the data for sprocket.
6. Click any point on the graphics drawing area (a duplicate of the sprocket appears; see Fig. 16-44).
7. Click Cancel in the Insert Components PropertyManager.

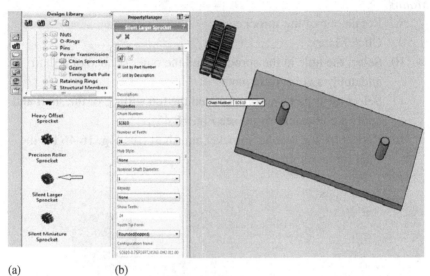

(a)   (b)
Figure 6-43 Chain Sprockets tool and Sprocket PropertyManager

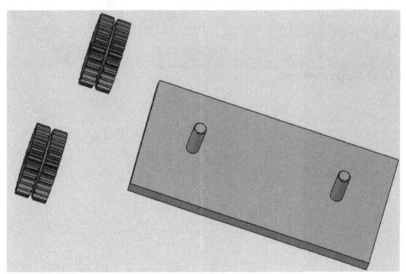

Figure 16-44 Sprockets, support plate, and shafts

Mating

8. Rotate one of the sprockets.

9. Click Mates.

10. Select the hole in the sprocket and the outer surface of a shaft and apply a concentric mate.

11. Select the top of the sprocket and the top surface of the shaft and apply a coincident mate (see Fig. 16-45).

12. Repeat for the second sprocket and shaft (see Fig. 16-46 for the final sub-assembly).

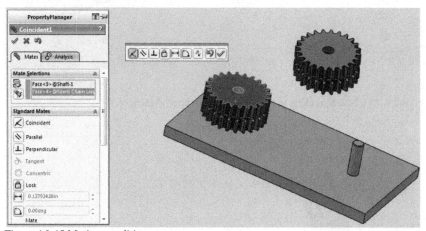

Figure 16-45 Mating conditions

Figure 16-46 Sub-assembly of support plate, shafts, and sprockets

Inserting the chain

13. Click Insert > Assembly Feature > Belt/Chain (see Fig. 16-47 for the Belt/Chain insertion tool).

14. Click the top surface of a tooth on the left sprocket and the top surface of a tooth on the right sprocket to define the Belt Members (see Fig. 16-48).

15. In the Properties rollout, check Use belt thickness and enter 0.5 in.

16. Check Engage belt, and Create belt part. A chain is created.

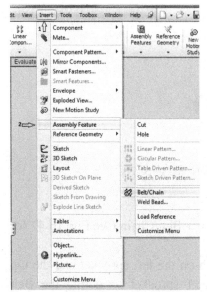

Figure 16-47 Belt/Chain insertion tool

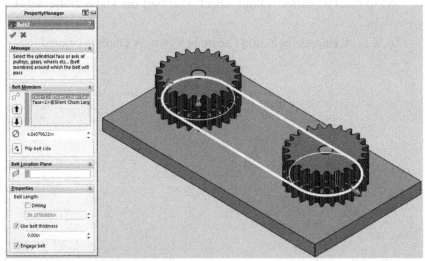

Figure 16-48 Belt Members defined

17. Expand Belt1 in the FeatureManager.

18. Right-click [Belt1^...]<1>->? and click the edit part icon (see Fig. 16-49).

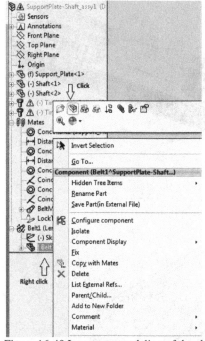

Figure 16-49 In-context modeling of the chain

19. Click any segment of the chain.
20. Click the sketch icon to enter sketch mode.
21. Click Features > Extrude Boss/Base. The Extrude PropertyManager is displayed as shown in Fig. 16-50.
22. In the Direction 1 rollout, select Mid Plane.
23. Set the extrusion distance to 1.00 in.
24. Check the box for Thin Feature and set the thickness to 0.50 in.
25. Click OK.
26. Click Edit Component to exit editing the assembly. Figure 16-51 shows the chain-and-sprocket assembly model.

Figure 16-50 Extrude PropertyManager

Figure 16-51 Chain-and-sprocket assembly model

Figure 16-52 shows the FeatureManager for the chain-and-sprocket assembly model.

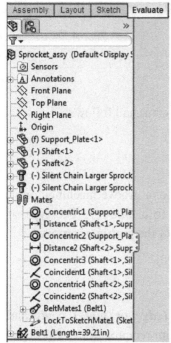

Figure 16-52 FeatureManager for the chain-and-sprocket assembly model

16.5.3 *Animation*

1. Click the Motion Study1 icon at the bottom right corner of the graphics window.
2. Click the motor icon.
3. Click one of the sprockets as the component.
4. Set the motion speed as required (e.g. 100 rpm).
5. Click OK.

16.6 Further Reading

http://en.wikipedia.org/wiki/Gear
http://en.wikipedia.org/wiki/Belt_(mechanical)
http://en.wikipedia.org/wiki/Chain_drive

Chapter 17

Cam Design

Objectives:

When you complete this chapter you will:

- Have learnt how to design cams
- Understand the relationship between cams and followers

17.1 Introduction

A *cam* is a mechanical device having a profile or groove machined on it, which gives an irregular or special motion to a *follower*. The type of follower and its motion depend on the shape of the profile or groove.

17.2 Types of Cams

Cams fall into two main classes, *radial (edge* or *plate) cams* and *cylindrical cams*. The follower of a radial cam reciprocates or oscillates in a plane perpendicular to the cam axis, whilst with a cylindrical cam the follower moves parallel to the cam axis.

17.3 Types of Followers

The *knife edge* or *point follower* is the simplest type of follower. It is not often used as it wears rapidly but it has the advantage that the cam profile can have any shape.

With the *roller follower* the rate of wear is reduced, but the profile of the cam must not have any concave portions with a radius smaller than the roller radius.

The *flat follower* is sometimes used but the cam profile must have no concave portions.

17.4 Creating Cams using the Traditional Method

The traditional method of creating cam profiles is to define a displacement diagram and then transfer the displacement diagram to a *base circle*.

What is a base circle? A base circle is a circle passing through the nearest approach of the follower to the cam center.

17.5 Creating Cams in SolidWorks

SolidWorks creates cams by utilizing existing cam templates. There are templates for *circular* and *linear* cams and *internal* and *external* cams. The templates allow the designer to work directly on the cam profile and eliminate the need for a displacement diagram. The SolidWorks Toolbox must be available before we can access the Cams Tool. Therefore, the first step is to Add-In the Toolbox.

17.5.1 *Problem definition*

Let us illustrate how to create a cam in SolidWorks with a circular cam having a 4.00-in base circle and a profile that rises 0.5 in over 90° using harmonic motion, dwells for 180°, falls 0.50 in over 45° using harmonic motions, and dwells for 45°.

17.5.2 *SolidWorks Toolbox add-ins*

1. Open SolidWorks.
2. Open the model file.
3. Click the Add-Ins tool (see Fig. 17-1).
4. Check SolidWorks Toolbox and SolidWorks Toolbox Browser (see Fig. 17-2).
5. Click OK (Toolbox tool is added).

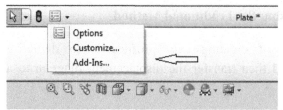

Figure 17-1 Add-Ins tool

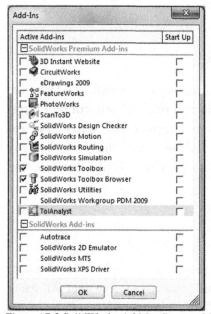

Figure 17-2 SolidWorks Add-Ins PropertyManager

17.5.3 *To access the SolidWorks Cams tool*

To access the Cams tool:

1. Create a new part document.
2. Select the front plane.
3. Click the toolbox icon from the top of the CommandManager.

4. Click the Cams tool from the menu (see Fig. 17-3). The Cam-Circular toolbox will be automatically displayed (see Fig. 17-4).

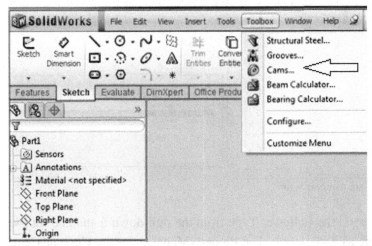

Figure 17-3 Cams tool

The Cam-Circular toolbox has three tabs: Setup, Motion, and Creation.

Cam-circular setup
1. Click the List button on the Setup tab of the Cam-Circular toolbox (see left side of Fig. 17-4). The Favorites dialog will be displayed (see right side of Fig. 17-4). This dialog lists the cam templates that can be used to create different cams.
2. Select Sample 2 - Inch Circular.
3. Click Load.

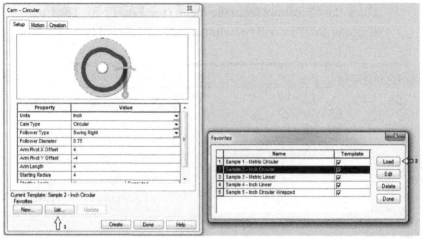

Figure 17-4 Cam-Circular Setup

4. Expand the Follower Type from the pull-down menu and set the type as Translating. The PropertyManager for the Circular-Cam, Translating-Follower appears (see Fig. 17-5).
5. Define the Properties required for the Setup of the cam.

The base circle of the cam has a diameter of 4.00 in (radius of 2.00 in) and the follower has a diameter of 0.50 in (radius of 0.25 in). This means that the Starting Radius is 2.25 in (2.00 + 0.25). Therefore, in the Property rollout on the Setup tab enter the following:

Units: Inch
Cam Type: Circular
Follower Type: Translating
Follower Diameter: 0.50
Starting Radius: 2.25
Starting Angle: 0
Rotation Direction: Clockwise

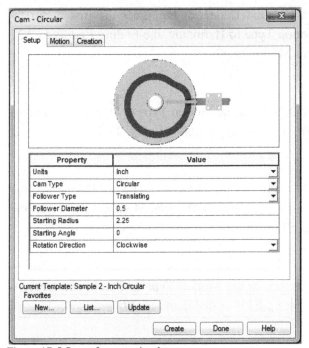

Figure 17-5 Setup for cam circular

Cam-circular motion

From the problem definition, the motion of the circular cam has four sectors. It rises 0.5 in over 90° using harmonic motion, dwells for 180°, falls 0.50 in over 45° using harmonic motions, and dwells for 45°. This information is utilized to define the motion.

6. Select the Motion tab in the Cam-Circular toolbox.
7. Click Add to display the Motion Creation Details dialog (see Fig. 17-6).

8. In the Motion Creation Details dialog, enter the first sector by setting the Motion Type to Harmonic, the Ending Radius to 2.25 in, and Degrees Motion to 90°.

9. Click OK.

10. Click Add.

11. For the second sector, set the Motion Type to Dwell and Degrees Motion to 180° (see the upper dialog in Fig. 17-7).

12. Click OK.

13. Click Add.

14. For the third sector, set the Motion Type to Harmonic, the Ending Radius to 2.25 in, and Degrees Motion to 45° (see the middle dialog in Fig. 17-7).

15. Click OK.

16. Click Add.

17. For the final sector, set the Motion Type to Dwell, and Degrees Motion to 45° (see the lower dialog in Fig. 17-7).

18. Click OK.

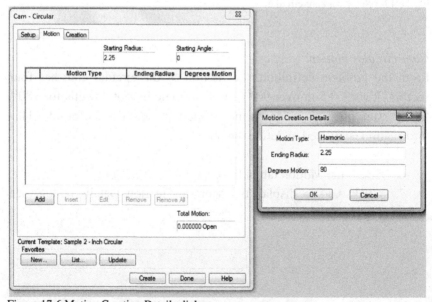

Figure 17-6 Motion Creation Details dialog

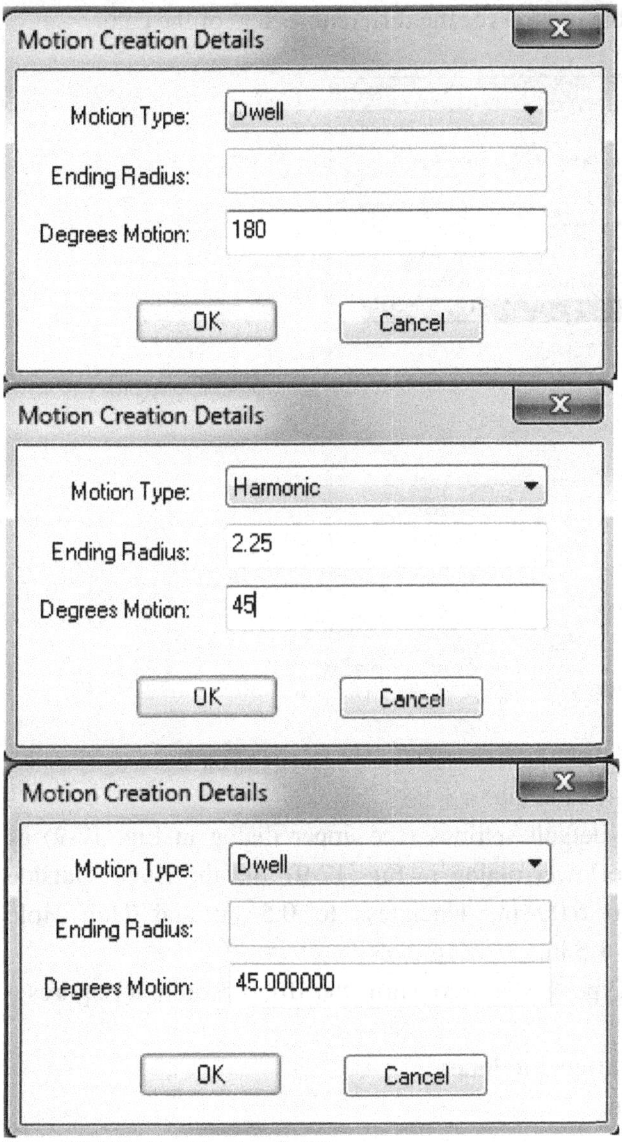

Figure 17-7 Motion Creation Details

Figure 17-8 shows the motion for the different sectors of the cam.

Figure 17-8 Motion summary

Cam-circular creation

19. Select the Creation tab.
20. Modify the default settings (see upper dialog in Fig. 17-9) as follows (see lower dialog in Fig. 17-9). Set the Blank Outside Diameter to 6.00 in, Thickness to 0.5 in; and Thru Hole Diameter to 1.5 in.
21. Set Track Type & Depth to Thru (the Blind condition requires a value).
22. Set Track Surfaces to Inner.
23. Click Create.
24. Click Done.

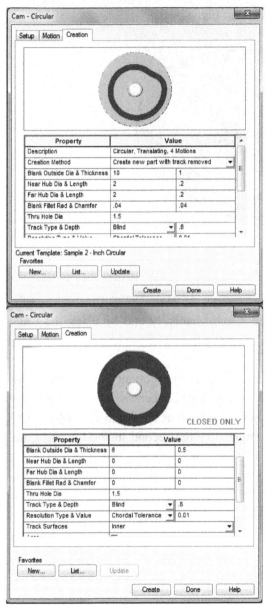

Figure 17-9 Creation tab for cam-circular

The final cam model is displayed in Fig. 17-10.

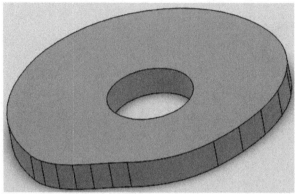

Figure 17-10 Final cam model

17.6 Creating a Hub

There are two ways to create a hub: manually and by using the Cam-Circular dialog.

In the manual method, the cam model is first obtained as shown in Fig. 17-10. Then using the Sketch tool, a circle concentric with the circle that defines the shaft is sketched, and extruded, as shown in Fig. 17-11. This is not an efficient method compared to the second method, using Cam-Circular dialog.

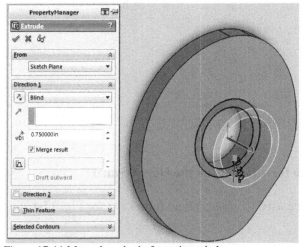

Figure 17-11 Manual method of creating a hub

17.7 Creating a Hub using the Cam-Circular Dialog

This is easily achieved by defining the following fields in Fig. 17-12:

Near Hub Diameter = 1.5 in

Near Hub Length = 0.75 in

The hub is created as shown in Fig. 17-13.

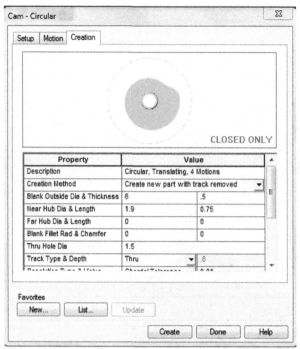

Figure 17-12 Hub definition in Creation tab for cam-circular

17.8 Creating a Hole for a Key using the Hole Wizard

The Hole Wizard can be used to create a hole for a key (see Fig. 17-13).

Type
1. Click the Type tab.
2. Click the Hole Wizard.
3. Select ANSI Inch Standard.
4. For Type, select the Tapered Hole.
5. For Size, select ¼-20.

Position
6. Click the Position tab.
7. Use the Smart Dimension tool to dimension the position of the center of the hole (see Fig. 17-14).

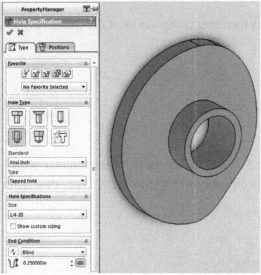

Figure 17-13 Creating a hole for a key using the Hole Wizard

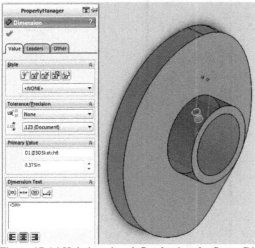

Figure 17-14 Hole location defined using the Smart Dimension tool

17.9   Cam Shaft Assembly

The following sections design the other components of the cam shaft assembly.

17.9.1 *Spring*

Section 6.2.1.1 gives the steps involved in spring design.

The helix path diameter is 0.5 in and the diameter of the circular profile is 0.125 in. The other specifications for the helix are shown in Fig. 17-15.

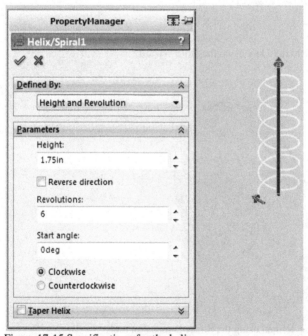

Figure 17-15 Specifications for the helix

17.9.2 *Cam follower bracket*

The cam follower bracket is modeled by creating a U-profile 1.00 in by 0.875 in by 1.00 in (see Fig. 17-16). The thickness is 0.13 in and it is extruded in the Mid Plane direction by 0.5 in (see Fig. 17-17).

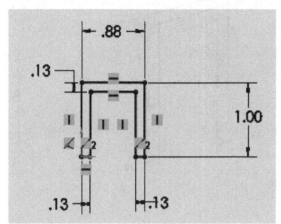

Figure 17-16 Profile for cam follower bracket

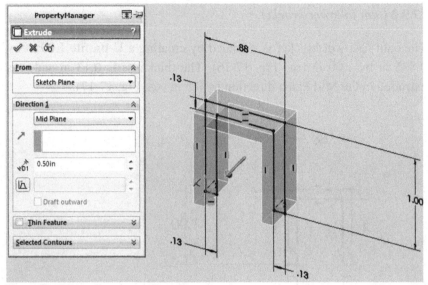

Figure 17-17 Extruding the profile

Face filleting

The two arms of the U, Face<1> and Face<2>, are filleting using the Tangent Propagation option (see Fig. 17-18).

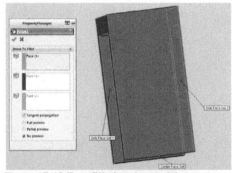

Figure 17-18 Face filleting the model

Holes of diameter 0.25 in are cut into the top and sides. The center of the hole in the top face coincides with the center of that face, while the centers of the holes in the sides coincide with the center used to generate the fillets, that is, 0.25 in from each side. The cam follower bracket model is shown in Fig. 17-19.

Figure 17-19 Cam follower bracket model

17.9.3 *Cam bracket*

The cam bracket is modeled by creating a U-profile 3.00 in by 8.5 in by 4.00 in (see Fig. 17-20). The thickness is 0.25 in and it is extruded in the Mid Plane direction by 6.00 in (see Fig. 17-21).

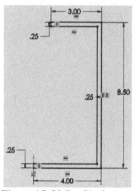

Figure 17-20 Profile for cam bracket

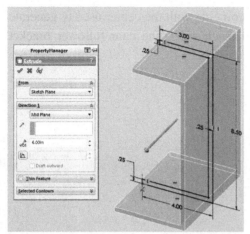

Figure 17-21 Extrusion of profile for cam bracket

17.9.4 Roller

The roller is modeled by creating a circle of diameter 0.50 in, and the hole is a concentric circle of diameter 0.25 in. Both circles are extruded through 0.5-in (see Fig. 17-22).

Figure 17-22 Roller model

17.9.5 Cam shaft

The cam shaft is modeled by creating a circle of diameter 1.5 in and extruded in the Mid Plane direction by 2.5 in (see Fig. 17-23).

628

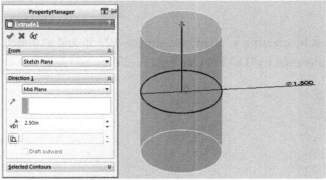

Figure 17-23 Cam shaft model

17.9.6 *Handle*

The handle is modeled by creating a circle of diameter 0.25 in and extruded in the Mid Plane direction by 2.75 in (see Fig. 17-24).

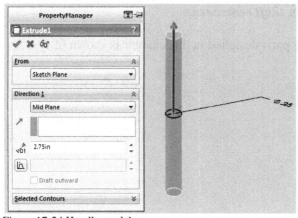

Figure 17-24 Handle model

17.9.7 *Pin*

The pin is modeled by creating a circle of diameter 0.25 in and extruded in the Mid Plane direction by 0.875 in (see Fig. 17-25).

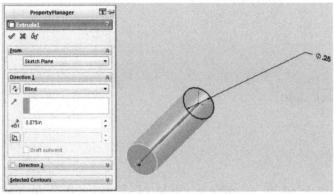

Figure 17-25 Pin model

17.9.8 *Assembly of cam shaft components*

The assembly of all the parts designed in this chapter is shown in Fig. 17-26.

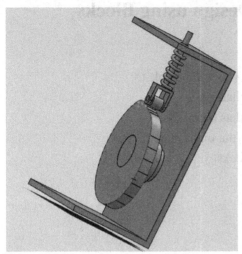

Figure 17-26 Assembly of cam and follower

Exercises

1. Assemble the follower bracket, roller, handle, and pin. Name this the cam sub-assembly.
2. Assemble the cam sub-assembly, cam, cam shaft, and spring. Name this the cam assembly.

All files for this assembly are available in the textbook's resources database.

Chapter 18

Mechanism Design using Blocks

Objectives:

When you complete this chapter you will be able to:
- Create sketches of a mechanism
- Save the sketches created as block files
- Insert a block into the layout environment
- Apply relations to the blocks
- Convert blocks into parts

18.1 Introduction

A block is a set of entities grouped together as a single entity. Blocks are used to create complex mechanisms as sketches and to check their functionality before being developed into complex 3D models.

18.2 Blocks Toolbar

The Blocks toolbar, shown in Fig. 18-1, is used to control the sketched entities of the blocks.

Figure 18-1 The Blocks toolbar

18.3 Problem Description

The different views of the parts of a reciprocating mechanism with required dimensions are shown in Figs 18-2, 18-3, and 18-4. We will create the sketches and save them as SolidWorks blocks. We will convert the blocks to parts and assemble the reciprocating mechanism.

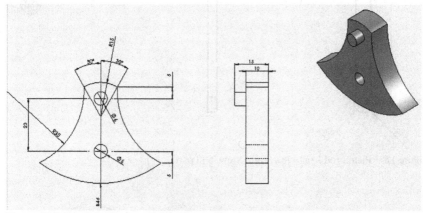

Figure 18-2 Crank front view, right view, and part

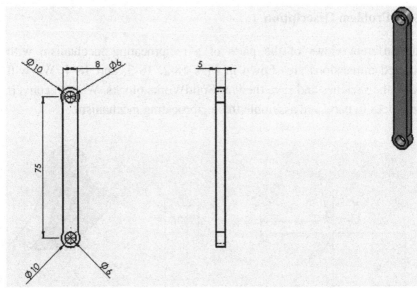

Figure 18-3 Piston rod front view, right view, and part

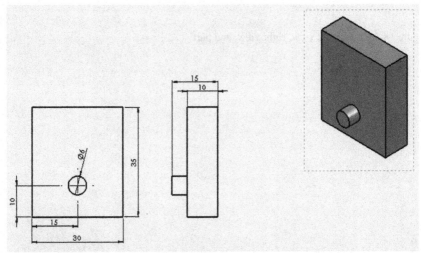

Figure 18-4 Piston tank front view, right view, and part

18.4   Creating Sketches of a Mechanism

Crank

1. Start a new SolidWorks part document.
2. Select the front plane.
3. Sketch the crank profile shown in Fig. 18-5.

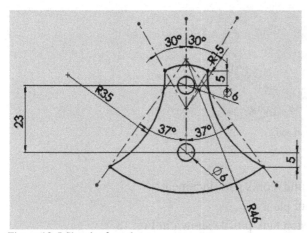

Figure 18-5 Sketch of crank

Piston rod

4. Start a new SolidWorks part document.
5. Select the front plane.
6. Sketch the piston rod profile shown in Fig. 18-6.

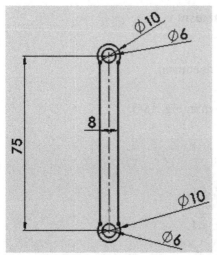

Figure 18-6 Sketch of piston rod

Piston tank

7. Start a new SolidWorks part document.
8. Select the front plane.
9. Sketch the piston tank profile shown in Fig. 18-7.

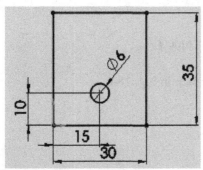

Figure 18-7 Sketch of piston tank

18.5 Saving the Sketches as Different Block Files

1. Save the sketch files as SolidWorks blocks.

18.6 Inserting the Block into the Layout Environment

1. Start a new SolidWorks assembly document.
2. Click the Create Layout button from the Begin Assembly PropertyManager (see Fig. 18-8). The Layout Command-Manager is displayed, as shown in Fig. 18-9.
3. Click Insert Block from the Layout CommandManager (see Fig. 18-9).

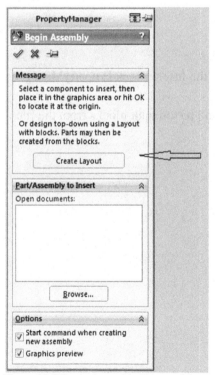

Figure 18-8 Begin Assembly PropertyManager

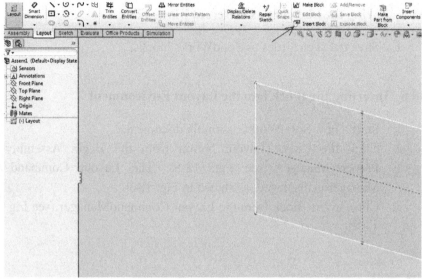

Figure 18-9 Layout CommandManager

4. Click the Browse button from the Insert Block PropertyManager (see Fig. 18-10).

5. Select the crank, piston rod, and piston tank blocks from the file (see Fig. 18-10).

6. Click OK.

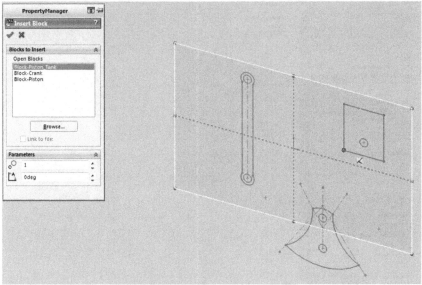

Figure 18-10 Opening the blocks for the crank, piston rod, and piston tank

18.7 Applying Relations to the Blocks

1. Click Add Relation from the CommandManager (see Fig. 18-11, arrow labeled 1). The Add Relations PropertyManager appears.
2. Select the origin and the center of the lower circle of the crank (see Fig. 18-11, arrows labeled 2 and 3).
3. Select the Coincident relation. Fig. 18-12 shows the crank located at the origin.

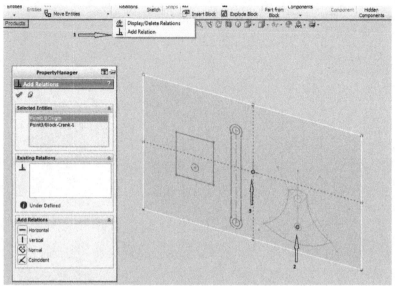

Figure 18-11 Add Relations PropertyManager for the crank

4. Click Add Relation from the CommandManager.
5. Select the center of the circle on the piston tank and the center of the upper circle of the piston rod (see Fig. 18-12).
6. Select the Coincident relation. Fig. 18-13 shows the piston tank and piston rod attached.

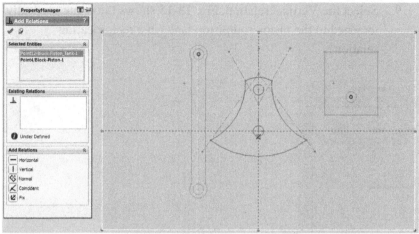

Figure 18-12 Add Relations PropertyManager for piston rod and piston tank

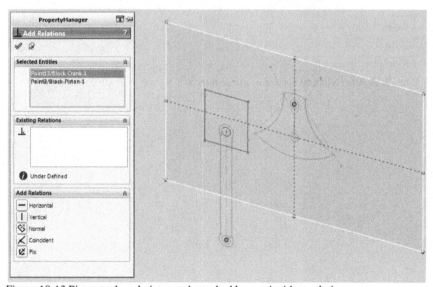

Figure 18-13 Piston tank and piston rod attached by a coincident relation

7. Click Add Relation from the CommandManager.
8. Select the center of the upper circle of the crank and the center of the lower circle of the piston rod (see Fig. 18-14).

9. Select the Vertical relation. Fig. 18-15 shows the crank vertically attached to the piston rod/piston tank (in a reciprocating relation).

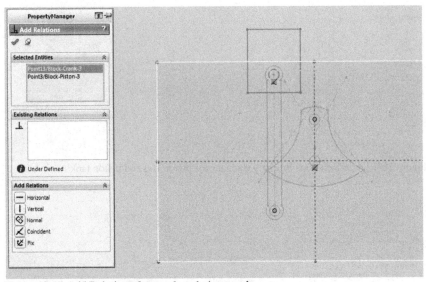

Figure 18-14 Add Relations for crank and piston rod

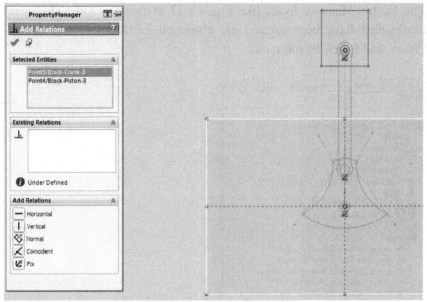

Figure 18-15 Crank vertically attached to piston rod/piston tank

Now, when the crank is rotated, the piston tank will reciprocate vertically due to the relations defined (see Fig. 18-16).

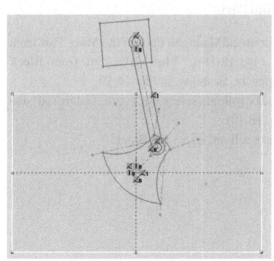

Figure 18-16 Crank, piston rod, and piston tank properly connected

643

The FeatureManager looks like Fig. 18-17 at this juncture. The three blocks that have been created are displayed as: Block-Crank, Block-Piston, and Block-Piston-tank.

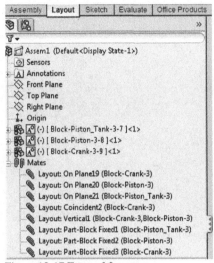

Figure 18-17 FeatureManager

18.8 Converting Blocks into Parts

1. From the Layout CommandManager, choose the Make Part from Block option (see Fig. 18-18). The Make Part from Block PropertyManager appears, as shown in Fig. 18.19.
2. In the Selected Blocks rollout, select the crank, piston rod, and piston tank (see Fig. 18.19).
3. In the Selected Blocks rollout, select On Block.

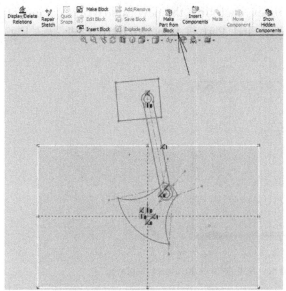

Figure 18-18 Make Part from Block option

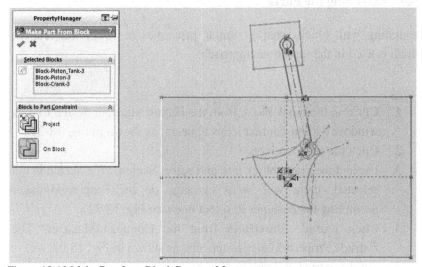

Figure 18-19 Make Part from Block PropertyManager

4. Click OK three times, each time the new SolidWorks document dialog is displayed and Part is selected by default (see Fig. 18-20).

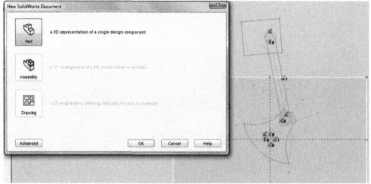

Figure 18-20 Click OK for the new SolidWorks document displayed

18.9 Extruding the Parts

Designing with blocks shares similar principles to in-context editing, which is used in the top-down approach.

Crank

1. Click on the crank block from the FeatureManager design tree. A window with in-context icons appears, as shown in Fig. 18-21.
2. Click the edit part icon.
3. Expand the crank block tree and select Sketch1. If a sketch is not selected, there will be a message on the PropertyManager prompting the designer to select one (see Fig. 18-22).
4. Click Extrude Base/Boss from the CommandManager. The Extrude PropertyManager appears, as shown in Fig. 18-23.
5. Set the extrusion depth to 10 mm.

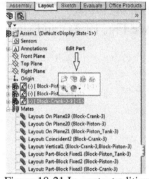

Figure 18-21 In-context editing

Figure 18-22 Extrude PropertyManager prompting user to select a sketch

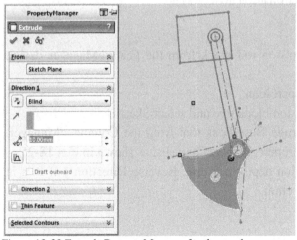

Figure 18-23 Extrude PropertyManager for the crank

Creating a boss to coincide with the upper circle on the crank

6. Click the right-hand side of the crank (see Fig. 18-24) and start sketch mode.
7. Click the circle.
8. Click Convert Entities to extract this circle.
9. Click the Extrude Base/Bose tool.
10. Set the extrusion depth to 15 mm.

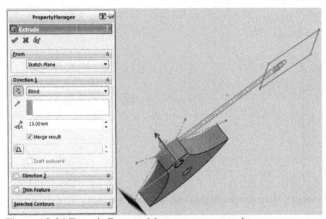

Figure 18-24 Extrude PropertyManager to create a boss

Piston rod

11. Click on the piston rod block from the FeatureManager design tree.
12. Click the edit part icon.
13. Expand the piston block tree and select Sketch1.
14. Click the Extrude Base/Boss tool from the CommandManager. The Extrude PropertyManager appears, as shown in Fig. 18-25.
15. Set the extrusion depth to 5 mm. Reverse Direction 1 if it is in the wrong direction.

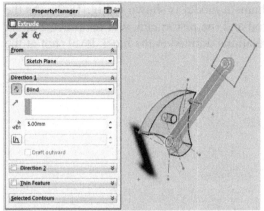

Figure 18-25 Extrude PropertyManager for the piston rod

Piston tank

16. Click on the piston block from the FeatureManager design tree.
17. Click the edit part icon.
18. Expand the piston tank block tree and select Sketch1.
19. Click the Extrude Base/Boss tool from the CommandManager. The Extrude PropertyManager appears, as shown in Fig. 18-26.
20. Set the extrusion depth to 5 mm. Reverse Direction 1 if it is in the wrong direction.

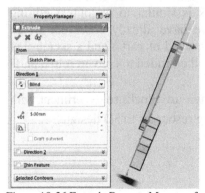

Figure 18-26 Extrude PropertyManager for the piston tank

The final reciprocating mechanism model is shown in Fig. 18-27. As the crank is rotated, the piston rod and the piston tank will reciprocate along the vertical axis due to the relations defined at the block level (not at the assembly level).

Figure 18-27 Final reciprocating mechanism model

Summary

The block design is now complete. We note that the mating conditions were imposed on the blocks rather than on the assembly of parts. This is the advantage of using blocks because we can sketch in 2D and add relations to 2D sketches, which are synonymous with mates in assembly mode. When blocks are used, the Layout tool is used instead of the Assembly tool. Designing with blocks is similar to in-context editing used in the top-down approach, as we have already observed. Block design creates an assembly of blocks (instead of parts) and a sketch that defines a block is accessed in-context and extruded to convert it to a part.

Industrial and engineering designers often use blocks to experiment with various designs in 2D before committing resources to 3D design.

Chapter 19

Threads and Fasteners

Objectives:

When you complete this chapter you will:

- Be able to model internal threads for nuts and other components
- Be able to model external threads for bolts, screws, and other similar components
- Understand the available SolidWorks Design Library Toolbox for creating threads and fasteners
- Be able to model nuts, bolts, screws, and other similar components using the Design Library Toolbox
- Be able to add bolts to an assembly of parts using the Smart Fasteners tool
- Be able to create retaining ring grooves on a cylindrical model

19.1 Threads and Fasteners

Nuts, bolts, screws, and other similar components are common fasteners for holding together two or more parts that are in contact. Generally, nuts have internal threads while bolts and screws have external threads. This chapter presents the basic concepts of modeling threads and fasteners using SolidWorks. When standard fasteners are used for engineering design, it is highly recommended that the SolidWorks Design Library of tools are used. The Appendix has useful information relating to sizes of threads and fasteners.

19.2 Internal and External Threads

The SolidWorks Hole Wizard is used to create internal threads in holes of a standard size for nuts. For standard fasteners, such as bolts and screws, the external thread is created using the SolidWorks Design Library of tools. When a hole diameter is not available in the Hole Wizard, or a bolt or screw dimension is not available in the Design

651

Library, then the user has to manually create the thread. This could be cumbersome in some cases.

19.3 Internal Thread Example

Problem description

We will create an internal thread for a 3/8-16 UNC hole through a 2 x 3 x 1.5 in block using the Hole Wizard.

Create the block
1. Start a new part document.
2. Select the top plane.
3. Create a rectangular profile, Sketch1, 2 in by 3 in (see Fig. 19-1).
4. Extrude along the Mid Plane by 1.5 in (see Fig. 19-2).

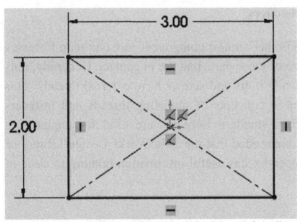

Figure 19-1 Sketch1 for creating a block for internal threading

Figure 19-2 Extruded base for internal threading

5. Click the top surface of the block (see Fig. 19-3).

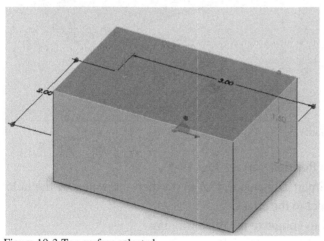

Figure 19-3 Top surface selected

6. Click Features > Hole Wizard (see Fig. 19-4).
7. Select the Type tab.
8. In the Hole Type rollout, set Standard to Ansi Inch, Type to Tapped hole, and Size to 3/8-16.

653

9. Set the End Condition to Through All.

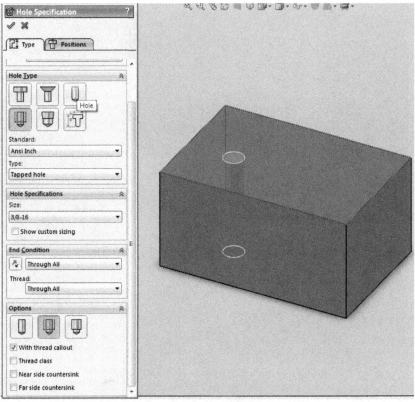

Figure 19-4 Hole Wizard

10. Select the Position tab (see Fig. 19-5).
11. Use the Smart Dimension tool to position the center of the hole with respect to the edges.
12. Click OK to complete the dimensioning.
13. Click OK from the Hole Position PropertyManager to conclude the hole creation process (see hole in Fig. 19-6).

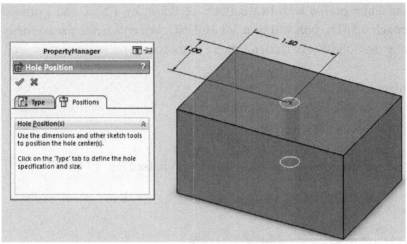

Figure 19-5 Locate the center of the hole using the Smart Dimension tool

Figure 19-6 Hole with thread created using the Hole Wizard

19.4 External Thread Example

Problem description

Three I-shaped blocks are to be fastened in an assembly using a hex bolt, two washers, and a nut. Each block is made of a 2.5 in by 3.5 in rectangular profile with two cut-outs of 0.5 in by 1.5 in, and extruded through 0.5. The bolt will be a 3/8-16 UNC. We will create the assembly using the SolidWorks Design Library of tools.

Create the block
1. Start a New Part document.
2. Select the top plane.
3. Create the profile, Sketch1 (see Fig. 19-7).
4. Extrude along the mid-plane by 1.5 in (see Fig. 19-8).

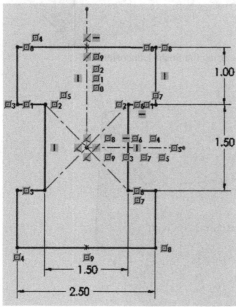

Figure 19-7 Profile for Sketch1

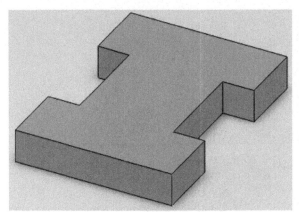

Figure 19-8 Extruded base for threading

Create a hole through the block

5. Click the top surface of the block (see Fig. 19-9).

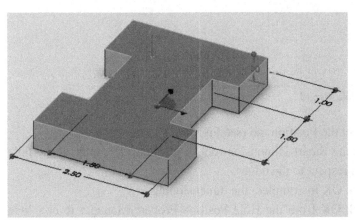

Figure 19-9 Top surface selected

6. Click Features > Hole Wizard (see Fig. 19-10).
7. Select the Type tab.
8. In the Hole Type rollout, set Standard to Ansi Inch, Type to Tapped hole, and Size to 3/8-16.
9. Set the End Condition to Through All.

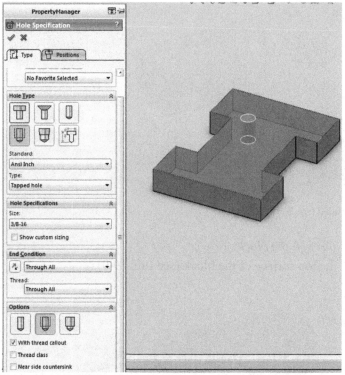

Figure 19-10 Hole Wizard

10. Select the Position tab (see Fig. 19-11).
11. Use the Smart Dimension tool to position the center of the hole with respect to the edges.
12. Click OK to complete the dimensioning.
13. Click OK from the Hole Position PropertyManager to conclude the hole creation process (see hole in Fig. 19-12).

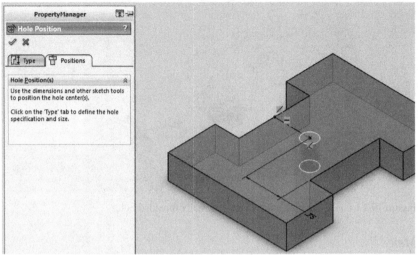

Figure 19-11 Locate the center of the hole using the Smart Dimension tool

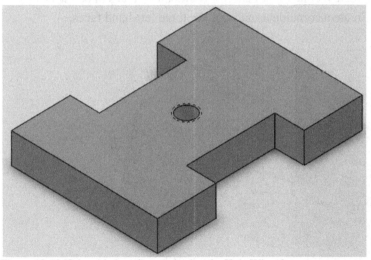

Figure 19-12 Hole with thread created using the Hole Wizard

Assemble the blocks

14. Create a new SolidWorks assembly document.

15. Open the I-shaped block.

16. Insert two more copies of the block (see Fig. 19-13).

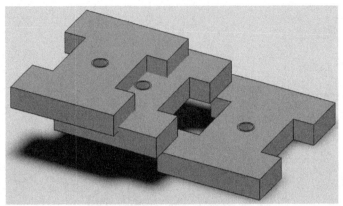

Figure 19-13 Three I-shaped blocks with internally threaded holes

Mates

17. Create a concentric mate for the three holes (see Fig. 19-14).
18. Create a coincident mate for each pair of faces in contact.
19. Create a coincident mate for the three left-hand faces.

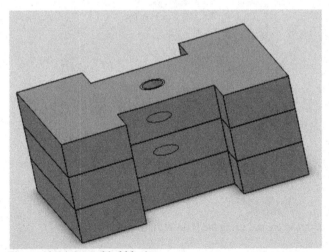

Figure 19-14 Assembled blocks

Design Library Toolbox

The Design Library Toolbox is an add-in, which must be enabled.
   20. Expand the Options toolbar and select Add-Ins.
   21. Check SolidWorks Toolbox (see Fig. 19-15).
   22. Check SolidWorks Toolbox Browser.

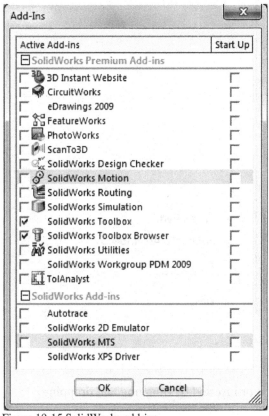

Figure 19-15 SolidWorks add-ins

   23. Click Design Library > Toolbox (see Fig. 19-16).

24. Click Ansi Inch.
25. Click Plain Washers (Type A).

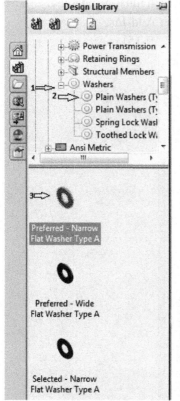

Figure 19-16 Design Library

Washer

Standard parts can be created or inserted by dragging and dropping.

Washer: drag-and-drop

26. Click and hold down the icon for Preferred - Narrow Flat Washer Type A and drag-and-drop it into the graphics window.

Washer: create part

27. Right-click the Preferred-Narrow Flat Washer Type A icon.

28. Click the Create Part option. A washer is created in the graphics window.

Sizing the washer

29. Size the washer by clicking the arrow to the right of the initial size and selecting a value.

30. Save the washer in a folder as a file.

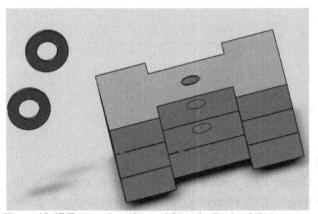

Figure 19-17 Two washers dragged from the Design Library

Add washers to the assembly

31. Insert two washers (see Fig. 19-17).

32. Create a concentric mate between the hole of the washer and the hole through the I-shaped feature.

Create a hex bolt

33. Click Design Library > Toolbox (see Fig. 19-18).

34. Click Ansi Inch.

35. Click Bolts and Screws > Hex Head.

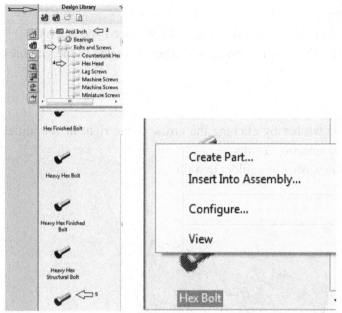

Figure 19-18 Creating a Hex Bolt

36. Right-click the Hex Bolt.
37. Click the Create Part option. The hex bolt is created in the graphics window. (Alternatively, insert the bolt using drag-and-drop.)
38. Size the hex bolt by clicking the arrow to the right of the initial size and selecting a value.
39. Save the hex bolt in a folder as a file.

Add the bolt to the assembly
40. Create a concentric mate between the holes of the washers, the hole through the I-shaped feature, and the hex bolt (see Fig. 19-19). The sub-assembly is shown in Fig. 19-20.

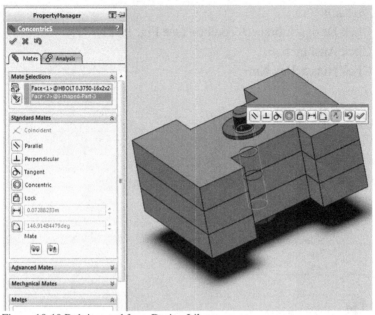

Figure 19-19 Bolt inserted from Design Library

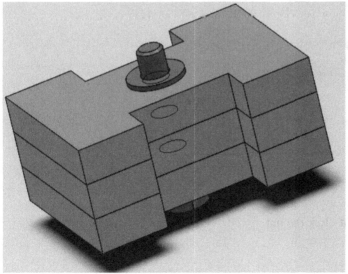

Figure 19-20 Sub-assembly with hex bolt

Create a hex nut

41. Click Design Library > Toolbox (see Fig. 19-21).
42. Click Ansi Inch.
43. Click Nuts > Hex Nuts.

Figure 19-21 Hex Jam Nut in Design Library

44. Right-click the nut.

45. Click the Create Part option. The nut is created in the graphics window.

46. Size the nut by clicking the arrow to the right of the initial size and selecting a value (see Fig. 19-22).

47. Save the nut in a folder as a file.

Figure 19-22 Nut inserted from Design Library

Add nut to the assembly

48. Create a concentric mate between the hole in the nut and the hex bolt (see Fig. 19-23). The final assembly is shown in Fig. 19-24.

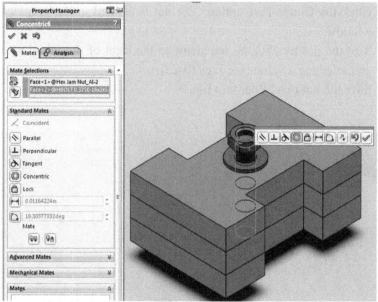

Figure 19-23 Mating the nut

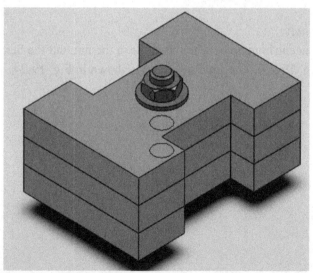

Figure 19-24 Final assembly of I-feature, washers, bolt, and nut

19.5 Smart Fasteners

The Smart Fasteners tool can be usd to automatically create the correct bolt for a hole in an assembly. Figure 19-25 shows a sub-assembly consisting of three I-shaped parts, two washers, and a nut. We will use the Smart Fasteners tool to add the appropriate bolt.

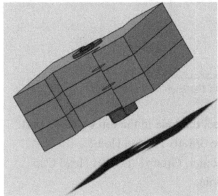

Figure 19-25 Sub-assembly consisting of three I-shaped parts, two washers, and a nut

1. Open the sub-assembly.
2. Click the Smart Fasteners tool located on the Assembly CommandManager. A warning dialog is shown (see Fig. 19-26).
3. Click OK to continue. The Smart Fasteners PropertyManager appears, as shown in Fig. 19-27.

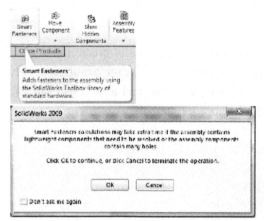

Figure 19-26 Warning dialog when the Smart Fasteners tool is selected

4. In the Selection rollout, select the hole in the top I-shaped part. The name of the hole will be 3/8-16 Tapped Hole1.
5. Click the Add button. A fastener, Group1 (Socket Head Cap Screw), will appear in the hole.

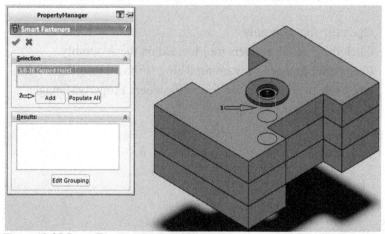

Figure 19-27 Smart Fasteners PropertyManager

6. In the Series Components rollout, right-click the Socket Head Cap Screw (see Fig. 19-28).

7. Click Change fastener type. The Smart Fastener window appears, as shown in Fig. 19-29(a).

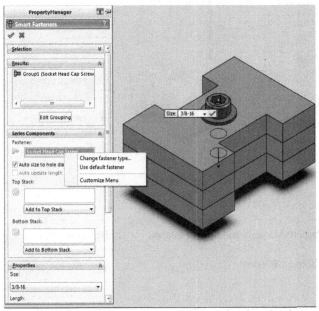

Figure 19-28 Smart Fasteners PropertyManager for changing fastener type

8. From the Fastener pull-down menu, select Hex Head.
9. Scroll down the Smart Fastener PropertyManager. In the Properties rollout, set Size to 3/8-16, Length to 2 in, and Thread Length to 2 in (see Fig. 19-29(b)).
10. Click OK on the Smart Fastener Window.
11. Click OK on the Smart Fastener PropertyManager. An appropriate bolt has now been added to the sub-assembly.

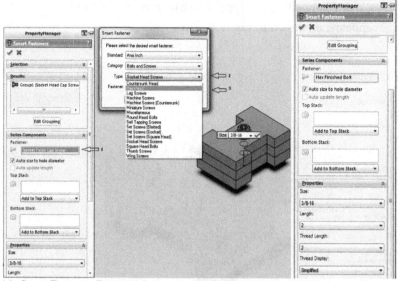

(a) Smart Fasteners PropertyManager and Smart Fastener Window;
(b) Bolt properties

Figure 19-29 Smart Fasteners Window

19.6 Set Screws

Set screws are fasteners used to hold parts (such as gears and pulleys) to rotating shafts or other moving objects to prevent slippage between the two objects. We manually created unthreaded set screws for the bench vise of Chapter 7.

As an illustration, we will add set screws to the collar of a shaft. First we will design the shaft and collar, and then add set screws to the collar of the shaft.

19.6.1 *Shaft*

A sketch of the shaft, the Revolve PropertyManager, and the shaft part are shown in Fig. 19-30.

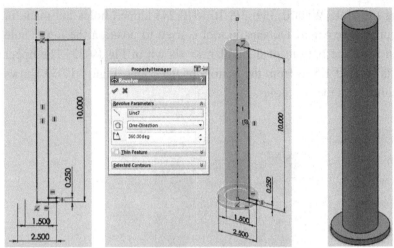

Figure 19-30 Sketch, Revolve PropertyManager, and the shaft part

19.6.2 *Collar*

The collar is created by extruding two concentric circles of diameters 1.5 in and 2.0 in, respectively, through 2.0 in. The Plane PropertyManager is used to define a plane to position the holes on the collar, as shown in Fig. 19-31.

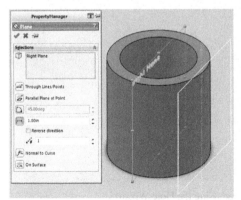

Figure 19-31Plane for positioning holes

Using the Hole Wizard, two 3/8-16 (#10-24) tapped holes are made in the collar. The Smart Dimension tool is used to position the lower hole 0.5 in from the bottom of the collar, as shown in Fig.19-32. The upper hole is located 1.5 in from the bottom of the collar. Figure 19-33 shows the collar with the two holes.

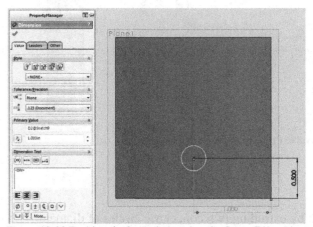

Figure 19-32 Position the lower hole using the Smart Dimension tool

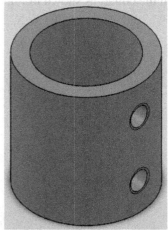

Figure 19-33 Collar with two holes inserted using the Hole Wizard

19.6.3 *Assembly of shaft and a collar*

Figure 19-34 shows the set screw selection in the Design Library Toolbox. The screw is dragged into the sub-assembly, as follows:

1. Open the collar and shaft sub-assembly.
2. Click Design Library > Toolbox > Ansi Inch > Bolts and Screws > Set Screws (Slotted).
3. Select the Slotted Set Screw Cup Point and drag it into the drawing window.
4. Define the size of the set screw as #10-24 and its length as 0.315 in.
5. Click OK.
6. Use the Mate tool to insert two set screws into the collar and shaft.
7. Save the assembly shown in Fig. 19-35.

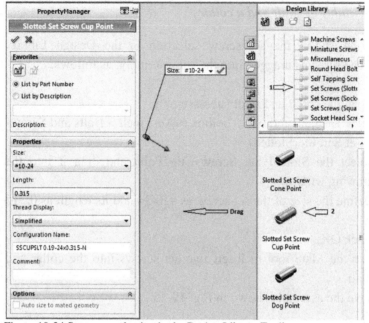

Figure 19-34 Set screw selection in the Design Library Toolbox

Figure 19-35 Assembly of shaft, collar, and set screws

19.7   O-Ring Grooves

We will create industry standard O-ring grooves on the part shown in Fig. 19-38.

A sketch of the part is shown in Fig. 19-36. The Revolve PropertyManager is shown in Fig. 19-37.

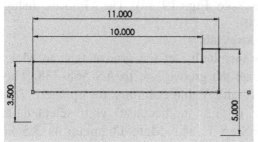

Figure 19-36 Sketch1

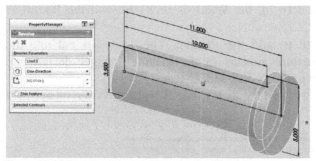

Figure 19-37 Revolve PropertyManager for Sketch1

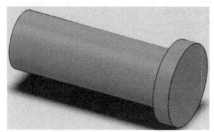

Figure 19-38 Revolved part

Creating the groove
1. Select a cylindrical face on the part where the groove is to be placed. By pre-selecting a cylindrical face, SolidWorks can

determine the diameter for the groove and suggest appropriate groove sizes.

2. Click Grooves on the SolidWorks Toolbox toolbar, or click Toolbox > Grooves (see Fig. 19-39). The Grooves dialog appears (see Fig. 19-40).

3. Select the O-Ring Grooves tab.

4. Set the standard to Ansi Inch, set the groove type to Male Static Groove, and set the groove size to AS 568-338. The fields in the Property and Value columns are updated. The Selected Diameter is 3.5 in because you selected a cylindrical face in Step 1. The Mate Diameter is 3.5 in because it is a reference value for the diameter of the non-grooved mating part that completes the seal. The Groove Diameter is 3.16 in, the Width is 0.281 in, and the Radius is 0.275.

5. Click Create to add the groove. The groove is cut into the model. A feature appears in the FeatureManager design tree called Groove for AS 568-338, O-Ring1.

6. Click Done to close the dialog. Fig. 19-41 shows the model with the groove.

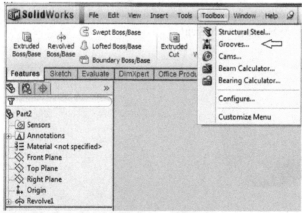

Figure 19-39 Toolbox > Grooves

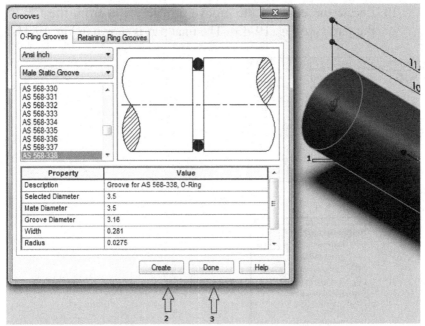

Figure 19-40 Grooves dialog

Figure 19-41 Model with O-ring groove

The next step is to locate the groove at the exact position required.

7. Right-click Sketch2 in the FeatureManager under Groove for AS 568-338, O-Ring1 Feature tree (see Fig. 19-42).
8. Click Edit Sketch to start sketch mode.
9. Position the part using Normal To tool.

10. Locate the edge of the groove profile 7.5 in from the top of the part (see Fig. 19-43). The final part is shown in Fig. 19-44.

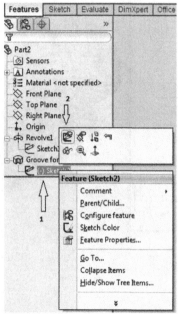

Figure 19-42 Selecting the groove

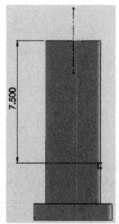

Figure 19-43 Changing the position of the groove

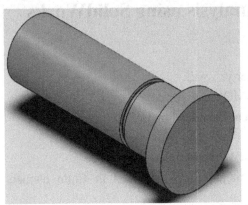

Figure 19-44 Final part with groove position adjusted

19.8 Retaining Ring Grooves

Industry standard retaining ring grooves can be created on your cylindrical model in the same way as for O-ring grooves. The Grooves dialog, shown in Fig. 19-45, has tabs for O-rings and retaining rings.

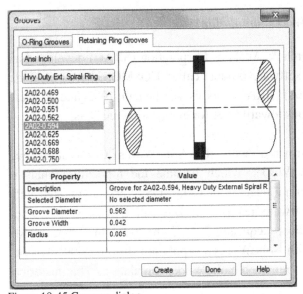

Figure 19-45 Grooves dialog

Chapter 20

Finite Element Analysis using SolidWorks

Objectives:

When you complete this chapter you will:

- Understand what COSMOS is used for
- Understand the historical trend leading to SolidWorks Simulation
- Be able to use the SolidWorks Simulation interface
- Understand the fundamental steps involved in finite element analysis (FEA)
- Understand static analysis with solid elements
- Understand the effect of mesh variations on analysis outputs
- Be able to use SolidWorks Simulation to solve stress analysis problems

20.1 Introduction to COSMOS/SolidWorks Simulation

20.1.1 *What is SolidWorks Simulation?*

The Structural Research & Analysis Corporation (SRAC) developed an engineering analysis software product called COSMOS, based on finite element analysis (FEA). SRAC was established in 1982 and it made significant contributions toward FEA for engineering analysis.

In 1995 SRAC partnered with the SolidWorks Corporation and developed COSMOS Works, which became the top-selling analysis solution. The commercial success of COSMOS Works integrated with SolidWorks CAD software resulted in Dassault Systemes, parent of SolidWorks Corporation, acquiring SRAC in 2001. In 2003, SRAC operations merged with SolidWorks Corporation. In the 2009 revision, COSMOS Works was renamed SolidWorks Simulation. This historical perspective is important.

SolidWorks Simulation is fully integrated into the SolidWorks Simulation CAD software. SolidWorks Simulation can be used to create and edit model geometry. SolidWorks Simulation is solid-driven, parametric, and feature-driven and runs on Windows.

There are a number of well-known commercially available FEA packages:

Software	Owner
ANSYS	ANSYS, Inc.
ABACUS	
SolidWorks Simulation/ COSMOS Works	Dassault Systemes
I-DEAS	UGS
Pro/MECHANICA	PTC

The functionalities, historical trends, and scope of the SolidWorks family of products are summarized in Table 20-1. A conceptual model of SolidWorks Simulation is shown in Fig. 20-1.

Table 20-1 The SolidWorks family

Functionality	Until 2008	From 2009	Scope
Stress Analysis: FEA-based Thermal Analysis: FEA-based	COSMOS Works[*]	SolidWorks Simulation[+]	Static, frequency, buckling, fatigue, drop test analysis, linear dynamic, nonlinear thermal analysis: temperature, temperature gradient, heat flow
Flow Analysis: FEA-based	COSMOS FloWorks	SolidWorks Flow Simulation	Fluid flow, heat transfer, forces
Motion Analysis	COSMOS Motion	SolidWorks Motion	Kinematic modeling/analysis of mechanisms
Animation	COSMOS Animation	SolidWorks Animation	Animation of modeled systems

[*] Designer and Professional versions

[+]SolidWorks SimulationXpress is an introductory version of SolidWorks Simulation

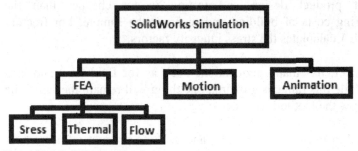

Figure 20-1 SolidWorks Simulation

20.1.2 *Product Development Cycle (PDC)*

In the industry, the product development cycle has the following steps:
1. Build model in a CAD system (Inventor, CATIA, Pro/E, SolidWorks, etc.).
2. Prototype the design.
3. Test the prototype in the field.
4. Evaluate the results of the field tests.
5. Modify the design based on the field test results.

The process continues until a satisfactory solution is reached.

FEA could be used to replace field tests in the PDC.

Advantages of analysis
- Reduced cost by simulation instead of tests
- Reduced development time
- Improved products

20.1.3 *What is finite element analysis?*

Finite Element Analysis (FEA) is a numerical method, usually modeled on a computer, which analyzes the stresses in a part. The results would otherwise be difficult to obtain. It can be used to predict the failure of a part or structure, due to unknown stresses, by showing problem areas and allowing designers to see all of the theoretical internal stresses. This

685

method of product design and testing is far cheaper than the manufacturing costs of building and testing each sample. For fracture analysis, FEA calculates the stress intensity factors.

FEA, however, has many applications such as for fluid flow and heat transfer. While this range is growing, one thing will remain the same: the theory of how the method works.

FEA is used in new product design, and existing product refinement. A company is able to verify that a proposed design will perform to the client's specifications prior to manufacturing or construction. It is used to ensure that a modified product or structure will meet its new specifications. In the case of structural failure, FEA may be used to help determine the design modifications necessary to overcome the problem.

There are generally two types of analysis that are used in industry: 2D modeling and 3D modeling. While 2D modeling is simple and only requires a relatively normal computer, it tends to yield less accurate results. 3D modeling, however, produces more accurate results while sacrificing the ability to run on all but the fastest computers. For each of these modeling schemes the programmer can insert numerous algorithms (functions) to make the system behave linearly or non-linearly. Linear systems are far less complex and generally do not take into account plastic deformation. Non-linear systems do account for plastic deformation, and many also are capable of testing a material all the way to fracture.

The stiffness of a member can be used to provide a simplified overview of the mathematical basis of finite element analysis. We begin by considering a simple member of original length L subject to an external axial deformation, ΔL. Force and deformation are related by

$$\Delta L = \frac{F.L}{A.E}$$

where E is the modulus of elasticity, F is the force, L is the length of the member, and the cross-sectional area is A.

The strain, which is the change in length divided by the original length, is defined as

$$\varepsilon = \frac{\Delta L}{L}$$

From the classical stress–strain relation, the stress can be determined as

$$\sigma = E\varepsilon$$

In other words, finite element analysis starts from a simple mathematical description of the deformation in a part due to some loading, then progresses to determine the strain, and finally finds the stress in the part. However, it should be noted that this stress formulation is only valid within the elastic region where stress is proportional to strain.

20.1.4 *How does finite element analysis work?*

In FEA, a part is divided into a number of simple elements:
- Rod
- Beam
- Plate/shell/composite
- Shear panel
- Solid
- Spring
- Mass
- Rigid element
- Viscous damping element

The most commonly used elements are solid, shell, and beam.

Solid elements

The majority of parts analyzed with FEA utilize 3D models, based on solid geometry, to define the boundaries of the part or assembly. The solid element is either a first-order tetrahedron (see Fig. 20-2), which has four flat faces and four vertices or a second-order tetrahedron (see Fig. 20-3), which has four flat faces and ten nodes that are the four vertices and the mid-points of its edges. Solid elements have three degrees of freedom per node consisting of three deformations.

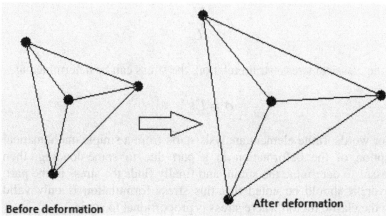

Before deformation   **After deformation**

Figure 20-2 First-order tetrahedral element before and after deformation

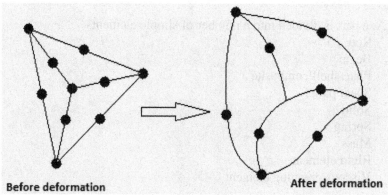

Before deformation   **After deformation**

Figure 20-3 Second-order tetrahedral element before and after deformation

Shell elements

Thin-walled parts, such as sheet metal parts, are analyzed using shell elements (see Fig. 20-4). Thin-walled parts are commonly found in tanks, beverage containers, plastic parts, thin-walled pressure vessels, etc. Shell elements have six degrees of freedom per node consisting of three deformations and three rotations.

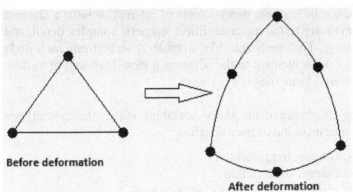

Before deformation

After deformation

Figure 20-4 Shell element before and after deformation

Beam elements

Beam or truss elements are commonly used in structural members since the number of nodes and elements is greatly reduced (see Fig. 20-5). A beam element should be used when the length-to-height ratio (*l/h*) is greater than or equal to 20:1.

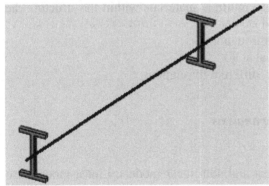

Figure 20-5 Beam element

FEA uses a complex system of points or nodes to make a grid called a mesh. This mesh is programmed to contain the material and structural properties that define how the structure will react under loading. Nodes are assigned at a density throughout the material depending on the anticipated stress levels of a particular area. Regions that will experience a high degree of stress usually have a higher node density than those which experience little or no stress. Points of interest include: a fracture point of a previously tested material, fillets, corners, complex detail, and high stress areas. The mesh acts like a spider's web; from each node there extends a mesh element to the adjacent nodes. This web of vectors carries the material properties of the object.

A wide range of objective functions (variables within the system) are available for minimization or maximization:

- Mass, volume, temperature
- Strain energy, stress, strain
- Force, displacement, velocity, acceleration

Different loading conditions may be applied to a system:
- Point, pressure, thermal, gravity, and centrifugal static loads
- Thermal loads derived from heat transfer analysis
- Enforced displacements
- Heat flux and convection
- Point, pressure and gravity dynamic loads

Many FEA programs can use multiple materials within the structure. he structure would be classified as:
- Isotropic, identical throughout
- Orthotropic, identical at 90°
- General anisotropic, different throughout

20.1.5 *Types of engineering analysis*

Structural
Structural analysis uses linear and non-linear models. Linear models use simple parameters and assume that the material is not plastically deformed. Non-linear models stress the material past its elastic capabilities. The stress in the material varies with the amount of deformation.

Vibration
Vibration analysis is used to test a material against random vibrations, shock, and impact. Each of these may act on the natural vibration

frequency of the material, which, in turn, may cause resonance and subsequent failure.

Fatigue

Fatigue analysis helps designers to predict the life of a material or structure by showing the effects of cyclic loading on a specimen. Such analysis can show the areas where crack propagation is most likely to occur. Failure due to fatigue may also show the damage tolerance of the material.

Heat Transfer

Heat Transfer analysis models the conductivity or thermal fluid dynamics of the material or structure. The heat transfer may be steady-state or transient. Steady-state transfer refers to constant thermo-properties in the material and yields linear heat diffusion.

20.1.6 *Principles of finite element analysis*

The methodology for FEA can be summarized as follows:
- Build the mathematical model using the CAD geometry (simplified if required), material properties, loads, restraints, types of analysis, etc. Different types of loads and connectors are shown in Tables 20-2 and 20-3.
- Build the finite element model by discretizing the mathematical model into solid elements, shell elements, beam elements, etc.
- Solve the finite element model (use the solver provided in SolidWorks Simulation).
- Analyze the results.

Table 20-2 Types of Loads

Structural Loads	Thermal Loads
Remote loads	Convection
Bearing loads	Radiation
Centrifugal loads	Conduction
Force	Temperature
Gravity	Heat flux
Pressure	Heat power
Shrink fit	

Table 20-3 Types of Connectors

Rigid connectors
Spring connectors
Pin connectors
Elastic support connectors

Build the mathematical model

The starting point for FEA using SolidWorks Simulation is the availability of a CAD model. If the part is complex, the model may need to be simplified (for example by removing fillets). Material properties are then assigned. The type of analysis is specified. Define the restraints and loads. This completes the mathematical model, as illustrated in Fig. 20-6.

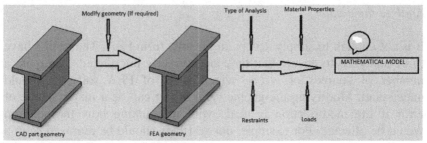

Figure 20-6 Mathematical model for FEA

Build the finite element model

Discretize the mathematical model using any of the following: solid elements, shell elements, or beam elements (see Fig. 20-7). This is also known as meshing the model. The geometry, loads, and restraints are all discretized and applied to the nodes of the elements. The elements are appropriately renumbered.

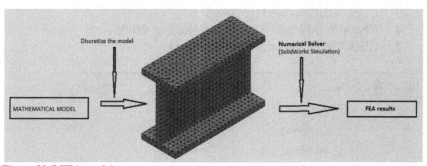

Figure 20-7 FEA model

Solve the finite element model

The finite element model is solved using a solver provided in SolidWorks Simulation (see Fig. 20-7). Solving FEA problems can take seconds for a simple model or hours for a complex model. Even for small problems, the number of nodes can run into thousands. Coarse elements yield inferior solutions compared to fine elements. The costs, in terms of computation time, are inversely proportional to the quality of the solution.

Analyze the results

It is not enough to simply accept any results from FEA. The results have to be analyzed to ensure that they are correctly interpreted. There are a number of sources of errors, which users of FEA software should understand. Modifying the geometry of a part can be a major source of error if the modification is made without thinking how the solution would be affected. For example, not all fillets should be removed from a part. Some fillets are necessary to reduce corner stresses. Discretizing the model is another area where errors could arise. The mesh size has a significant impact on the quality of the solution as previously discussed.

20.1.7 *SolidWorks Simulation add-ins*

SolidWorks Simulation is an add-in, which must be enabled:
1. Open SolidWorks.
2. Open a model file.
3. Click Add-Ins (see Fig. 20-8).
4. Check SolidWorks Simulation (see Fig. 20-9).
5. Click OK (Simulation tool is added).

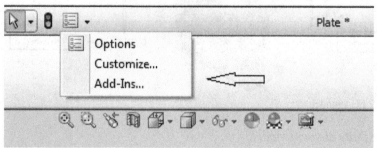

Figure 20-8 Add-Ins option

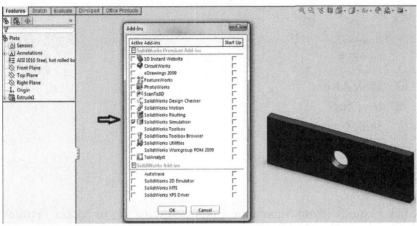

Figure 20-9 SolidWorks add-ins

20.1.8 *SolidWorks Simulation CommandManager*

The SolidWorks Simulation CommandManager has a number of advisors. These are found in the menus for Study, Fixtures, External Loads, Connections, Run, and Results, as shown in Fig. 20-10. A simulation advisor is a set of tools that guides you through the analysis process. The simulation advisor works with the SolidWorks Simulation interface by starting the appropriate PropertyManager and linking to online help topics for additional information.

Figure 20-10 SolidWorks Simulation CommandManager

Study Advisor

Click Study (from Simulation CommandManager) to access Study Advisor (see Fig. 20-11). The simulation advisor tab appears in the task pane. It recommends study types and outputs to expect. Study Advisor helps you define sensors and creates studies automatically.

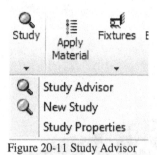

Figure 20-11 Study Advisor

Fixtures Advisor

Click Fixtures (from Simulation CommandManager) to access Fixtures Advisor (see Fig. 20-12). Fixtures Advisor defines internal interactions between bodies in the model. Fixtures are restraints applied to the model.

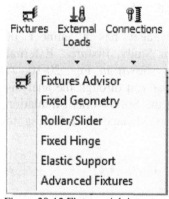

Figure 20-12 Fixtures Advisor

External Loads Advisor

Click External Loads (from Simulation CommandManager) to access External Loads Advisor (see Fig. 20-13). External Loads Advisor defines external interactions between the model and the environment. There are several types of external loads: force/torque, pressure, gravity, centrifugal force, bearing load, remote load/mass, distributed load, temperature, flow effects, thermal effects, etc.

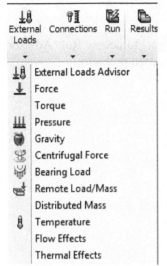

Figure 20-13 External Loads Advisor

Connections Advisor

Click Connections (from Simulation CommandManager) to access Connections Advisor (see Fig. 20-14). Connections Advisor suggests techniques for connecting components within an assembly model.

Figure 20-14 Connections Advisor

Run Advisor

Click Run (from Simulation CommandManager) to access Run Advisor (see Fig. 20-15). Run Advisor solves the simulation problem.

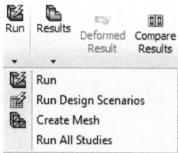

Figure 20-15 Run Advisor

Results Advisor

Click Results (from Simulation CommandManager) to access Results Advisor (see Fig. 20-16). It provides tips for interpreting and viewing the output of the simulation. Also, it helps to determine if frequency or buckling might be areas of concern.

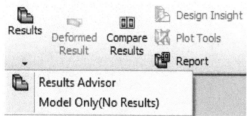

Figure 20-16 Results Advisor

Design Scenario

Click Design Scenario (from Simulation CommandManager) to access Design Scenario (see Fig. 20-17).

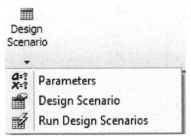

Figure 20-17 Design scenario

20.1.9 *SolidWorks Simulation toolbars*

Another way to access the functions for creating, solving, and analyzing a model is through the SolidWorks Simulation toolbars (see Fig. 20-18).

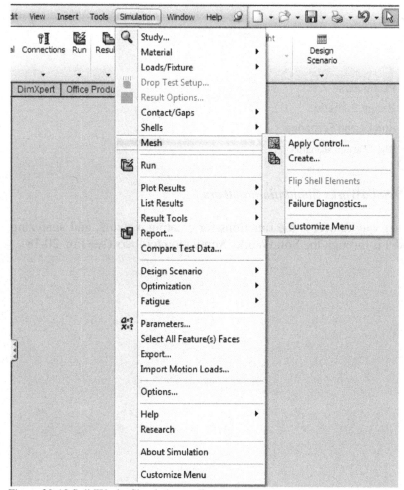

Figure 20-18 SolidWorks Simulation toolbars

20.1.10 *Starting a new study in SolidWorks Simulation*

1. Open a model file.
2. Click Simulation > New Study (see Fig. 20-19).

The Study PropertyManager has the following options, as shown in Fig. 20-20:
- Static (the default)

- Frequency
- Buckling
- Thermal
- Drop Test
- Fatigue
- Optimization
- Nonlinear
- Linear Dynamic
- Pressure Vessel Design

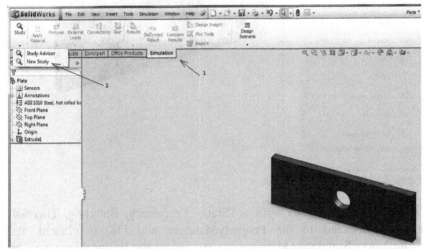

Figure 20-19 Simulation > New Study

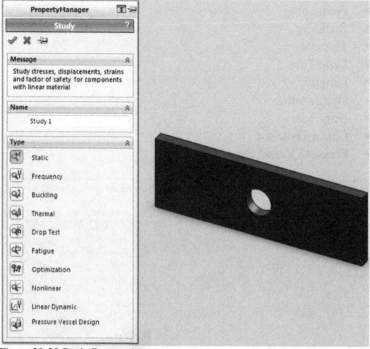

Figure 20-20 Study PropertyManager

When any of the study options (Static, Frequency, Buckling, Thermal, etc.) is selected in the PropertyManager and OK is clicked, the SolidWorks SimulationManager appears below the FeatureManager. The following options are roots of the SimulationManager:

- Connections
- Fixtures
- External Loads
- Mesh

20.1.11 *Basic SolidWorks Simulation steps*

The steps involved in SolidWorks Simulation for solving FEA problems (any of the study options) are summarized as follows:
1. Geometric preparation (if required).
2. Apply material to the model.
3. Define connections.

4. Define fixtures.
5. Define external loads.
6. Create the model mesh.
7. Run the model solution.
8. Analyze the results.

20.2 Finite Element Analysis of a Plate

Start a new study in SolidWorks Simulation:
1. Select Simulation > New Study.
2. Click OK (to use the default study option, Static).

We are now ready to define the analysis model. The user has to define the connections, fixtures, external loads, and mesh (see Fig. 20-21). If material has already been assigned to the part, then Material is not listed as one of the options. Applying materials during part design is the preferred approach, especially in cases where several parts make up an assembly.

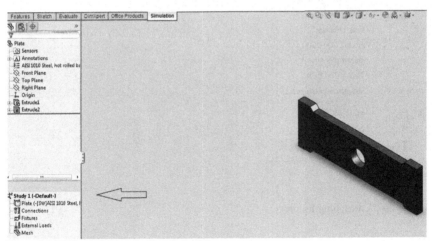

Figure 20-21 Study folders in SimulationManager

Defining Connections

3. Right-click the Connections folder and select the appropriate connections (see Fig. 21-22).

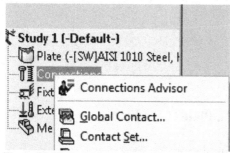

Figure 20-22 Assigning connections

Defining Fixtures

4. Right-click the Fixtures folder and select the appropriate fixtures (see Fig. 20-23).
5. Click Fixed Geometry (the preview of Fig. 20-24 appears).

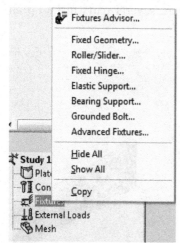

Figure 20-23 Assigning fixtures

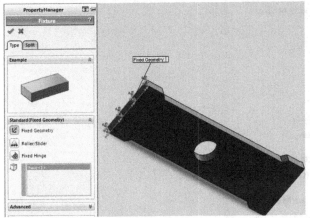

Figure 20-24 Fixtures preview for fixed geometry

Defining External Loads

6. Right-click the External Loads folder and select the appropriate external loads (see Fig. 20-25).
7. Click Force, a preview appears (see Fig. 20-26).
8. For the force, check Normal.
9. Select the face to apply the force, and enter a strength of 100 kN (for tensile forces, check Reverse direction).

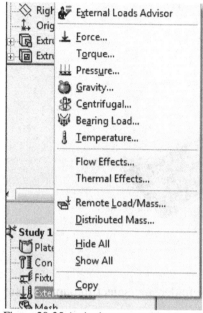

Figure 20-25 Assigning external loads

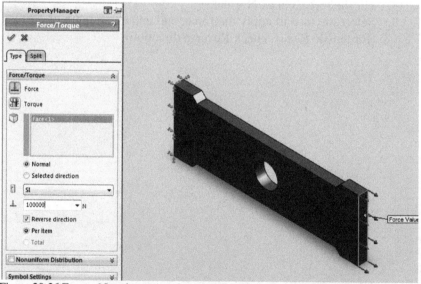

Figure 20-26 External Loads preview for normal force

Defining the mesh

10. Right-click the Mesh folder and select the appropriate mesh (see Fig. 20-27).
11. Click Create Mesh (the preview of Fig. 20-28 appears). Note that we can control the mesh density by moving the slider from Coarse to Fine. The element size (4.786 mm) and the element size tolerance (0.239 mm) are automatically established based on the geometric features of the SolidWorks model. The meshed model is shown in Fig. 20-29.

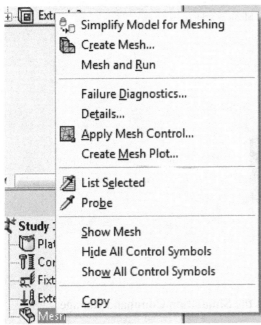

Figure 20-27 Assigning the mesh

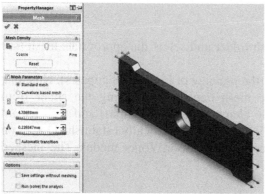

Figure 21-28 Mesh preview (Create Mesh is selected)

Figure 20-29 Meshed model

Running the model solution

12. Click Run Advisor in the Simulation CommandManager.

The results are shown (see Fig. 20-30) and three plots are automatically created in the Results folder:
- Stress1: von Mises stresses
- Displacement1: resultant stresses
- Strain1: equivalent strain

Figure 20-30 Model solution based on von Mises criterion

To show a plot, right-click on it and select Show (see Fig. 20-31 for the displacement plot).

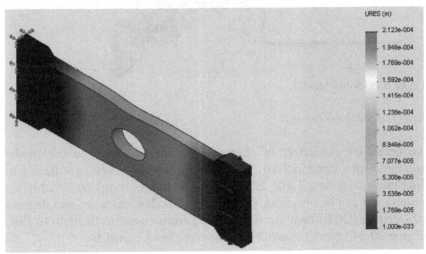

Figure 20-31 Model solution (displacement)

Saving analysis results

13. Right-click Results and select Solver Messages (see Fig. 20-32).
14. An analysis summary is shown, as in Fig. 20-33.
15. Click Save and OK (and give the path for saving the current solution).

709

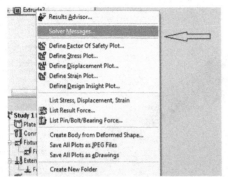

Figure 20-32 Accessing results for current simulation

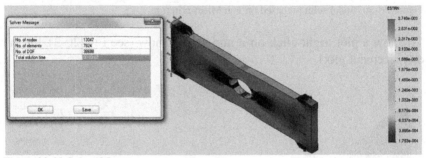

Figure 20-33 Solver Message

Effect of mesh refinement

To determine the effect of changing the mesh density on the model solution, we carried out an experiment, choosing coarse, medium and fine mesh densities. Table 20-4 shows the results based on von Misses criteria. As can be observed, the number of nodes, elements, and degrees of freedom (DOF) increase significantly from coarse to medium to fine. Figure 20-34 shows the results for the three mesh densities.

Table 20-4 Results for different mesh densities based on von Mises criterion

	Coarse	Medium	Fine
No. of nodes	2474	13047	76594
No. of elements	1318	7824	50012
No. of DOF	7287	38688	228225
Total solution time	00:00:01	00:00:02	00:00:08

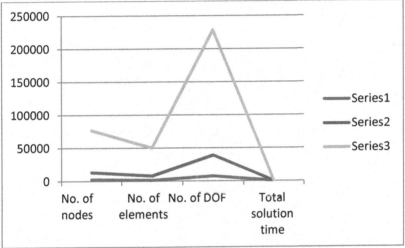

Figure 20-34 Results for three mesh densities (Series1=coarse; Series2=medium; Series3=fine)

Summary

When we run FEA software we must understand the results. For simple parts with classical solutions, it is good engineering practice to compare the FEA solution with a manual calculation. Once we have mastered FEA, then we can be more confident of our results when we solve complex, unfamiliar problems.

Exercises

1. Use the Probe tool to examine the stress around the central hole for the model presented in this chapter.
2. Apply the same load (100 kN) but this time, in a compressive manner. Plot the von Mises stress distribution and compare the stress around the central hole in the model. How does this differ from the distribution obtained in (1) above?

Appendix

A.1 Drill Sizes and Conversion Table

DRILL LETTER	DIAMETER (in)	DRILL NUMBER	DIAMETER (in)	DRILL NUMBER	DIAMETER (in)
A	.234	1	.228	31	.120
B	.238	2	.221	32	.16
C	.242	3	.213	33	.113
D	.246	4	.209	34	.111
E	.250	5	.2055	35	.110
F	.257	6	.204	36	.1065
G	.261	7	.201	37	.104
H	.266	8	.199	38	.1015
I	.272	9	.196	39	.0995
J	.277	10	.1935	40	.098
K	.281	11	.191	41	.096
L	.290	12	.189	42	.0935
M	.295	13	.185	43	.089
N	.302	14	.182	44	.086
O	.316	15	.180	45	.082
P	.323	16	.177	46	.081
Q	.332	17	.173	47	.0785
R	.339	18	.1695	48	.076
S	.348	19	.166	49	.073
T	.358	20	.161	50	.070
U	.368	21	.159	51	.067
V	.377	22	.157	52	.0635
W	.386	23	.154	53	.0595
X	.397	24	.152	54	.055
Y	.404	25	.1495	55	.052
Z	.413	26	.147	56	.0465
		27	.144	57	.043
		28	.1405	58	.042
		29	.136	59	.041
		30	.1285	60	.040

A.2 BA, BSF, and BSP Thread Sizes

THREAD SIZE	TPI*	MAJOR NOMINAL	EFFECTIVE DIA. NOMINAL	MINOR DIA. NOMINAL
6 BA	47.9	.1102	.0976	.0850
4 BA	38.5	.1417	.1262	.1106
2 BA	31.4	.1850	.1659	.1468
¼ BSF	26	.2500	.2254	.2008
9/32	26	.2812	.2566	.2320
5/16	22	.3125	.2834	.2543
3/8	20	.3750	.3430	.3110
7/16	18	.4375	.4019	.3663
½	16	.5000	.4600	.4200
9/16	16	.5625	.5225	.4825
5/8	14	.6250	.5793	.5336
11/16	14	.6875	.6418	.5961
¾	12	.7500	.6966	.6432
7/8	11	.8750	.8168	.7586
1	10	1.0000	.9360	.8720
1⅛	9	1.1250	1.0539	.9828
1¼	9	1.2500	1.1789	1.1078
1⅜	8	1.3750	1.2950	1.2150
1½	8	1.5000	1.4200	1.3400
⅛ *BSP*	28	.3830	.3601	.3372
¼	19	.5180	.4843	.4506
3/8	19	.6560	.6223	.5886
½	14	.8250	.7793	.7336
5/8	14	.9020	.8563	.8106
¾	14	1.0410	.9953	.9496
7/8	14	1.1890	1.1433	1.0976
1	11	1.3090	1.2508	1.1926
1¼	11	1.6500	1.5918	1.5336
1½	11	1.8820	1.8238	1.7656
1¾	11	2.1160	2.0578	1.9996
2	11	2.3470	2.2888	2.2306

*TPI: # Threads Per Inch

A.3 BSW Thread Sizes

NOMINAL DIAMETER	TPI*	MAJOR DIAMETER	EFFECTIVE DIAMETER	MINOR DIAMETER
1/8	40	.1250	.1090	.0930
3/16	24	.1875	.1608	.1341
¼	20	.2500	.2180	.1860
5/16	18	.3125	.2769	.2413
3/8	16	.3750	.3350	.2950
7/16	14	.4375	.3918	.3461
½	12	.5000	.4466	.3932
9/16	12	.5625	.5091	.4557
5/8	11	.6250	.5668	.5086
11/16	11	.6875	.6293	.5711
¾	10	.7500	.6860	.6220
7/8	9	.8750	.8039	.7328
1	8	1.0000	.9200	.8400
1⅛	7	1.1250	1.0335	.9420
1¼	7	1.2500	1.1585	1.0670
1½	6	1.3750	1.3933	1.2866
1¾	5	1.5000	1.6219	1.4938
2	4.5	2.0000	1.8577	1.7154
2¼	4	2.2500	2.0899	1.9298
2½	4	2.5000	2.3399	2.1798
2¾	3.5	2.7500	2.5670	2.3840
3	3.5	3.0000	2.8170	2.6340
3¼	3.23	3.2500	3.0530	2.8560
3½	3.25	3.5000	3.3030	3.1060
3¾	3	3.7500	3.5366	3.3232
4	3	4.0000	3.7866	3.5732

*TPI: # Threads Per Inch

A.4 Standard Hexagon Sizes

BRITISH HEXAGONS

NOMINAL ACROSS FLATS	ASSOCIATED SCREW THREAD BA & BSF	WIDTH ACROSS CORNERS APPROX. MAX
.193	6 BA	.220
.248	4 BA	.290
.324	2 BA	.370
.445	¼	.510
.525	5/16	.610
.600	3/8	.690
.710	7/16	.820
.820	½	.950
.920	9/16	1.060
1.010	5/8	1.170
1.100	11/16	1.270
1.200	¾	1.390
1.300	7/8	1.500
1.390	15/16	1.600
1.480	1	1.710
1.670	1⅛	1.930
1.860	1¼	2.150
2.050	1⅜	2.360
2.220	1½	2.560

UNIFIED HEXAGONS

NOMINAL ACROSS FLATS	ASSOCIATED SCREW THREAD UNF OR UNC	WIDTH ACROSS CORNERS APPROX. MAX
¼	#6 (.138)	.290
5/16	#8 (.164)	.360
11/32	#10 (.190)	.400
7/16	¼	.500
½	5/16	.580
9/16	3/8	.650
11/16	7/16	.790
¾	½	.860
7/8	9/16	1.010
15/16	5/8	1.080
1 1/16	¾	1.230
1¼	7/8	1.440
1 7/16	1	1.660

References

Pickup, F., and Parker, M. A., Engineering Drawing with Worked Examples 2, Hutchinson Educational, 1973.

Index

717